Internet of Things

Prospects in Networking and Communications – P-NetCom
Series Editor: Mohammad M. Banat

Internet of Things

A Hardware Development Perspective
Mohammad Ayoub Khan

For more information about this series, please visit: https://www.routledge.com/Prospects-in-Networking-and-Communications/book-series/NETCOM

Internet of Things

A Hardware Development Perspective

Edited by

Mohammad Ayoub Khan

CRC Press
Taylor & Francis Group
Boca Raton London New York

CRC Press is an imprint of the
Taylor & Francis Group, an **informa** business

First edition published 2022
by CRC Press
6000 Broken Sound Parkway NW, Suite 300, Boca Raton, FL 33487-2742

and by CRC Press
4 Park Square, Milton Park, Abingdon, Oxon, OX14 4RN

CRC Press is an imprint of Taylor & Francis Group, LLC

© 2022 selection and editorial matter, Mohammad Ayoub Khan; individual chapters, the contributors

Reasonable efforts have been made to publish reliable data and information, but the author and publisher cannot assume responsibility for the validity of all materials or the consequences of their use. The authors and publishers have attempted to trace the copyright holders of all material reproduced in this publication and apologize to copyright holders if permission to publish in this form has not been obtained. If any copyright material has not been acknowledged please write and let us know so we may rectify in any future reprint.

Except as permitted under U.S. Copyright Law, no part of this book may be reprinted, reproduced, transmitted, or utilized in any form by any electronic, mechanical, or other means, now known or hereafter invented, including photocopying, microfilming, and recording, or in any information storage or retrieval system, without written permission from the publishers.

For permission to photocopy or use material electronically from this work, access www.copyright.com or contact the Copyright Clearance Center, Inc. (CCC), 222 Rosewood Drive, Danvers, MA 01923, 978-750-8400. For works that are not available on CCC please contact mpkbookspermissions@tandf.co.uk

Trademark notice: Product or corporate names may be trademarks or registered trademarks and are used only for identification and explanation without intent to infringe.

ISBN: 978-0-367-64146-7 (hbk)
ISBN: 978-0-367-64148-1 (pbk)
ISBN: 978-1-003-12235-7 (ebk)

DOI: 10.1201/9781003122357

Typeset in Times
by KnowledgeWorks Global Ltd.

This work is dedicated to my wonderful parents, who always trusted me and allowed me do anything I wanted. I owe all I have today to God. This work is also dedicated to my wonderful wife, son, and little angel Amira Fatimah, who have been a huge source of encouragement for me to go farther in life.

Mohammad Ayoub Khan

Contents

Preface ix
About the Author xi
Contributors xiii

PART I Formal Design Flow for IoT Hardware

Chapter 1 Scalable Design and Processor Technology for IoT Applications 3

Mohammad Ayoub Khan and Amit Kumar

Chapter 2 Design Methods and Approaches for IoT Hardware 13

Ali Nezaratizadeh

Chapter 3 IoT Solution Reference Architectures 39

Vinay Chowdary, Tiyasa Bera, and Arpit Jain

PART II Simulation, Modeling, and Programming Framework

Chapter 4 Hardware Architecture of IoT and Wearable Devices 55

Manoj Sharma

Chapter 5 Cache Memory Design for the Internet of Things 75

Reeya Agrawal and Neetu Faujdar

Chapter 6 Investigation of Deep Learning Models for IoT Devices 107

Swagata Bhattacharya and Debotosh Bhattacharjee

PART III Communication Technologies and Trends

Chapter 7 Communication Technologies for M2M and IoT Domain 133

Manoj Kumar and Sushil Kumar

Chapter 8	Security Challenges and Solutions in IoT Networks for the Smart Cities	161
	A. Procopiou and T.M. Chen	
Chapter 9	Internet of Vehicles: Design, Architecture, and Security Challenges	205
	Abdullah Alharthi, Qiang Ni, and Richard Jiang	

PART IV Use Cases

Chapter 10	A Case Study on the Smart Streetlighting Solution Based on 6LoWPAN	223
	Manoj Kumar, Prashant Pandey, and Salil Jain	
Chapter 11	IoT-Enabled Real-Time Monitoring of Assembly Line Production	239
	Maneesh Tewari and Devaki Nandan	
Chapter 12	IoT-Enabled Hazardous Gas Leakage Detection System for Citizen's Safety	257
	Prerna Sharma and Latika Kharb	
Index		271

Preface

The Internet of Things (IoT) is the fastest growing technology that is being adapted by market and many industries to improve operational expenses, product life, and health. The IoT is a hot topic that combines hardware, embedded software, web services, and electronics to create cutting-edge devices that can be used in many applications like industry, retail, smart home, smart cities, and healthcare. However, there is no standard hardware for IoT. The IoT is based on the customized architecture and infrastructures to address needs in application-specific domains such as transportation, traffic, health, and environment. This book focuses on the hardware architecture, protocols, communication patterns, architectures, and interoperable issues important to IoT.

The book has 12 chapters. It starts with the fundamental of hardware and design flow for an IoT system. We have three chapters on these topics covering "Scalable Design and Processor Technology for IoT," "Design Methods and Approaches for IoT Hardware," and "IoT Solution Reference Architectures." The second part of the book consists of three chapters about the "Hardware Architecture of IoT and Wearable Devices," "Cache Memory Design for the Internet of Things," and "Investigation of Deep Learning Models for IoT Devices." In the third part, we have three chapters on "Communication Technologies for M2M and IoT Domain," "Security Challenges and Solutions in IoT Networks for the Smart Cities," and "Internet of Vehicles: Design, Architecture, and Security Challenges." The last part contains the case studies on real-life system that includes "A Case Study on the Smart Streetlighting Solution Based on 6LoWPAN," "An IoT-Enabled Real-Time Monitoring of Assembly Line Production," and "IoT-Enabled Hazardous Gas Leakage Detection System for Citizen's Safety."

This book will be a good resource for industry practitioners, research scholars, and academicians to develop new ideas for the IoT.

About the Author

Mohammad Ayoub Khan is working as a research associate professor at University of Bisha, Saudi Arabia with interests in Internet of Things, blockchain, RFID, wireless sensors networks, ad hoc network, smart cities, industrial IoT, and signal processing, NFC, routing in network-on-chip, real time and embedded systems. He has more than 14 years of experience in his research area. He has published more than 70 research papers and books in the reputed journals and international IEEE conferences. He is contributing to the research community by various volunteer activities. He has served as the conference chair in various reputed IEEE/Springer international conferences. He is a senior member of professional bodies of IEEE, ACM, ISTE, and EURASIP society. He may be reached at ayoub.khan@ieee.org.

Contributors

Reeya Agrawal
Department of Electronics & Communication Engineering
GLA University
Mathura, India

Abdullah Alharthi
School of Computing & Communication
Lancaster University
Lancaster, United Kingdom

Tiyasa Bera
Department of Electrical and Electronics
University of Petroleum and Energy Studies
Dehradun, India

Debotosh Bhattacharjee
Department of Computer Science & Engineering
Jadavpur University
Kolkata, India

Swagata Bhattacharya
Department of Electronics and Communication Engineering
Guru Nanak Institute of Technology
Kolkata, India

T.M. Chen
Department of Electrical Engineering
University of London
London, United Kingdom

Vinay Chowdary
Department of Electrical and Electronics
University of Petroleum and Energy Studies
Dehradun, India

Neetu Faujdar
Department of Computer Engineering & Applications
GLA University
Mathura, India

Arpit Jain
Department of Electrical and Electronics
University of Petroleum and Energy Studies
Dehradun, India

Richard Jiang
School of Computing & Communication
Lancaster University
Lancaster, United Kingdom

Mohammad Ayoub Khan
Department of Information Technology
College of Computing and Information Technology
University of Bisha
Bisha, Kingdom of Saudi Arabia

Latika Kharb
Department of Information Technology
Jagan Institute of Management Studies (JIMS)
Delhi, India

Amit Kumar
School of Computational Science and Engineering
Georgia Institute of Technology
Atlanta, Georgia

Sushil Kumar
Telecommunication Engineering Center (TEC)
New Delhi, India

Manoj Kumar
STMicroelectronics
Noida, India

Devaki Nandan
Industrial and Production Engineering
College of Technology
Pantnagar, India

Ali Nezaratizadeh
Department of Electrical Engineering
Shahid Rajaee Teacher Training
 University
Tehran, Iran

Qiang Ni
School of Computing &
 Communication
Lancaster University
Lancaster, United Kingdom

Prashant Pandey
STMicroelectronics
Noida, India

A. Procopiou
Department of Computer Science
Centre for Software Reliability
University of London
London, United Kingdom

Salil Jain
STMicroelectronics
Noida, India

Manoj Sharma
Department of Electronics and
 Communications Engineering
Bharati Vidyapeeth's College of
 Engineering
New Delhi, India

Prerna Sharma
Department of Information
 Technology
Jagan Institute of Management
 Studies (JIMS)
Delhi, India

Maneesh Tewari
Department of Industrial and
 Production Engineering
College of Technology
Pantnagar, India

Part I

Formal Design Flow for IoT Hardware

1 Scalable Design and Processor Technology for IoT Applications

Mohammad Ayoub Khan
University of Bisha
Bisha, Kingdom of Saudi Arabia

Amit Kumar
Georgia Institute of Technology
Atlanta, Georgia

CONTENTS

1.1 Introduction 3
1.2 High-Level IoT Characteristics and Architectures.......................................4
    1.2.1 Heterogeneity..5
    1.2.2 Scalability 6
    1.2.3 Real Time 6
    1.2.4 Intelligence in IoT Devices 7
    1.2.5 Complexity..7
1.3 Applications of IoT 7
    1.3.1 Smart Home Appliances 8
    1.3.2 Smart Agriculture...8
    1.3.3 Smart Healthcare 8
    1.3.4 Smart Cities 9
    1.3.5 Smart Industry 9
    1.3.6 Smart Retail..9
1.4 Reconfiguration of IoT Processors..9
1.5 Conclusion 10
References... 10

1.1 INTRODUCTION

Internet of Things (IoT) has extended in various sectors of life including healthcare, industry, security, and communication [1–4]. IoT and cloud-based technologies have led to the growth of connected devices, products with a wide array of functions, and increased computing capacity [5–7]. In turn, this has increased the level of sophistication of product design for engineers. The complexity of choosing resources for one of these devices becomes especially apparent. The issue is that engineers not only have to satisfy an increasing variety of apparently contradictory design requirements such as low power operation and high performance but also a variety of other processing options have to do so. This is true that all the network devices fulfill certain basic and universal functions that processing architectures can handle for general purpose. It is now necessary for IoT to be able to perform certain specific tasks including machine learning, speech or gesture recognition, and security [8–10]. These requirements have led to designers turning to a rising and changing accelerator class. The demands of the market make the designer's work even harder such as shorter cycles and lower development costs, making the process selection even more important.

Design practices of IoT are evolving day by day. It used to be that developers just looked at far-reaching processes, but now we monitor them in real time [11–13]. This has led to an improvement in the size of the IoT network. This can lead to several challenges when it comes to network paths for cloud servers with IoT devices that rely on Internet access: high latencies, low bandwidths, and reduced response time. The trends have brought new topologies in IoT networks, including fog computing. The deployment of cloud components at the edge of the network eliminates latencies while avoiding bandwidth bottlenecks. The edge networks and fog computing require high-performance computing, storage, and networking services in order to achieve these objectives. There are three major challenges in the design of IoT processors:

1. Design scalability and the reliability of the IoT processors.
2. Architectural flexibility and configurability.
3. Design IoT processors with minimum latency and highest throughput.

Though Intel offers a broad variety of processors that help designers accommodate all design scenarios in terms of scaling hardware and software. Most processors in this class also have an integrated graphics processing unit (GPU) that improves overall computing performance. Intel® has Quark™, Core™, Atom®, and Xeon® processor families that can be used in IoT. These four families of processors provide excellent speed, low power consumption, and improved bandwidth without increasing latency. With the growth in IoT, the architectural complexities have also increased. These problems can be solved with an Intel processor. The specifications of the IoT industry for various levels of power are requiring a wide variety of processes of power. Businesses recognize the necessity of computing and analytics being located in the cloud (Figure 1.1) [14, 15].

Edge devices are now being used to process information and react more quickly and precisely because of advanced microprocessors. Because of this rising need, edge computing is applied in new applications. Table 1.1 presents the summary of different processors, manufacturers, technologies, and power consumption.

1.2 HIGH-LEVEL IoT CHARACTERISTICS AND ARCHITECTURES

The characteristics of IoT demand new designs and optimization methods for processors that are deployed in IoT devices. The main characteristics of IoT are classified as shown in Figure 1.2 [16].

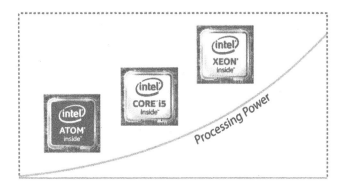

FIGURE 1.1 Intel processor family suitable for IoT [14, 15].

TABLE 1.1
Summary of Processors for IoT

Processors	Manufacturers	Technology Support	Power Consumption
Quark™, Atom® and Xeon®	Intel Corporation	Quark SoC X1000—Clanton/ Galileo Gen 2, Processor E3900 series, Intel Celeron™ Processor N3350, Intel Xeon Processor E5-2600 v4 Product Family with Intel C612 Chipset	Low
MediaTek MT3620	Microsoft worked with MediaTek	ARMCortex-M4F core, 500-MHz Cortex-A7 apps processor, Azure Sphere, Cloud	High
ON Semi RSL10	ON Semiconductor	ARM Cortex-M3	Low
ETA Compute Tensai	ETA	ARM Cortex-M3	Ultra-low
Microchip SAM R34/35	Microchip Technology STMicroelectronics	SAM R34/35 and LoRa transceiver	Low
NXP i.MX-RT600	NXP Semiconductors	ARM Cortex-M33,	Low
Renesas Electronics RZ/A2M	Renesas	DRP, ARM Cortex-A9	Low
Ambiq Apollo3 Blue	Ambiq	Cortex-M4, BLE 5radio	Low
STM32H7	STMicroelectronics	ARM Cortex M-7	Low
Quectel BG96	Quectel Wireless Communications	QUALCOMM MDM9206, NB-IoT	Low

1.2.1 Heterogeneity

Heterogeneity refers to the multiple architectures, hardware, protocols, and platforms of IoT devices. The IoT has a high degree of heterogeneity that features many different kinds of devices, applications, and contexts [16]. The processor must be

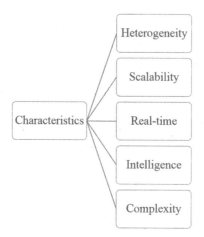

FIGURE 1.2 Characteristics of IoT [16].

designed to achieve heterogeneity at chip level by incorporating different cores. Also, the IoT devices must be able to integrate and communicate seamlessly.

1.2.2 Scalability

The scalability refers to the ability to scale the performance in the proportion of a number of devices that are increased. The scalability in IoT can be in sensors, networks, and the cloud. The sensor devices collect raw data from the environment that may include a variety of data such as temperature, pressure of water flow, or humidity readings. As the amount of data increases, it requires to stack multiple sensors which shall be scalable. The network of sensors must be scalable to achieve throughput and robustness. Next, scalability at cloud is applied since the data are sent to cloud servers for analysis and actions. The cloud should be able to serve multiple requests from these sensors while maintaining the turnaround time and throughput. The scalability must be efficient in terms of cost, energy, and area. An example is illustrated in Figure 1.3 to understand the relationship of tools and libraries for IoT deployment [15].

The Intel sensor library has about 300 commercial sensors that can work with many developments' environment and OS with real-time performance [15]. A company like Honeywell and many IoT giants are continuously contributing by the addition of many sensors. Intel Corporation has integrated many hardware and software that include sensor drivers, Intel boards, datasheets, and protocols thus saving large amount of time and costs for developers [15].

1.2.3 Real Time

All the IoT devices work in a real-time environment; therefore, they have many real-time constraints. One of the most important constraints is to meet the deadline for the task. The task execution must adhere to stringent deadlines. Therefore, the

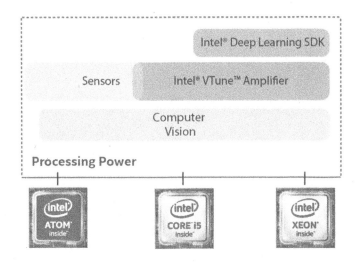

FIGURE 1.3 Tools and technologies for IoT deployment [15].

processors shall be able to dynamically determine and adhere to deadlines based on the application characteristics and quality of service (QoS).

1.2.4 Intelligence in IoT Devices

Nowadays, every device needs intelligence akin to human thinking. The objective is to minimize the reliance on human intervention during the data acquisition from the environment. The data reception and processing from the environment must be autonomous to take the right action. The processor should be able to adapt to execution scenarios and exceptions [17, 18].

1.2.5 Complexity

Complexity refers to the high degree of management of large numbers of heterogeneous architectures and applications [19]. The architecture should be able to execute a variety of applications. In IoT, many applications are processor-centric while some are memory-centric. The processor has to manage the behaviors of such applications.

1.3 APPLICATIONS OF IoT

The IoT has a huge potential for a variety of application domains such as healthcare, logistics, agriculture, smart home, and environment. The main goal of the IoT is to equip edge devices with sufficient computing resources that can perform fast computations, otherwise shall be assigned or transferred to a high-performance device. Based on the many applications, the IoT application can be broadly classified as shown in Figure 1.4.

There are many more applications of IoT, but we have discussed some of them in the next subsections.

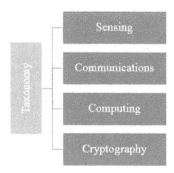

FIGURE 1.4 Taxonomy of IoT.

1.3.1 SMART HOME APPLIANCES

This area is not new but with the advent of IoT it has geared up. In home automation, the system monitors home appliances, entertainment system, lighting, and climate, and takes the right decision based on the predefined parameters. Home automation makes our life easier by following all of the predefined instructions. Presently, there are many products for home automation from companies such as Nest, Ecobee, Ring, and August, to name a few, which will become household brands and are planning to deliver a never seen before experience [19].

1.3.2 SMART AGRICULTURE

This one is the less addressed area in IoT. The rapid growth in world's population has increased the demand for food supply. The state agencies are helping farmers to use advanced techniques and research to increase food production [19]. Farming operations are generally based in remote locations and require a lot of effort to maintain the large number of livestock; all of this can be monitored by the IoT system. The IoT can change the way farmers operate on a daily basis to make it smoother. However, as noted earlier, smart agriculture is unaddressed and the idea is yet to reach a large-scale attention. Therefore, smart farming has a lot of potential to become an important application field, specifically in the agricultural-product exporting countries [19].

1.3.3 SMART HEALTHCARE

Smart healthcare is a very important application area of IoT that has a variety of applications in medical field [19]. Smart healthcare helps in remote patient monitoring tools to advance and smart sensors to equipment integration in medical health which has the potential to improve the way medical practitioners deliver care [19]. This enables patients to spend more time interacting with the doctors that can enhance patient engagement and satisfaction level. The IoT brings new tools and technology in the ecosystem that helps in creating better healthcare experience [1].

1.3.4 SMART CITIES

Smart city refers to the city which uses smart technologies to enhance the citizen's experience. Smart city is an application of IoT producing interest among world's population such as smart surveillance, automated transportation, children tracking, smarter energy management systems, water distribution, urban security and environmental monitoring, smart garbage/waste management using CrAN (Crowd-Associated Network), and green and clean environment [19]. IoT can helps in solving some of the critical issues such as pollution, traffic congestion, and shortage of energy supplies. The GPS system and other smart sensors can send information from drivers' cell phones so that smart stopping arrangements can decide if the parking areas are available or reachable and make an ongoing stopping map [19]. Based on the information received, it is easy to find out a parking space faster rather than aimlessly driving around [19].

1.3.5 SMART INDUSTRY

Smart industry is one of the areas that can improve productivity, supply chain, and logistics. The industrial Internet connects machines and devices in industries such as electricity generation, oil, gas, and healthcare [19]. The smart industry takes appropriate actions beforehand where unplanned downtime and system failures can result in life-threatening situations. In a smart industry, industrial devices embedded with the IoT tend to encompass gadgets such as monitoring sensors and actuators.

1.3.6 SMART RETAIL

The IoT has a huge potential in the retail sector as well. The IoT can provide a new opportunity for retailers to connect with the customers that enhances the in-store experience. Smartphones can be the one way for retailers to remain connected with their consumers even out of store [19]. The retailer can track the consumer's path through a store and can improve the store's layout and place premium products in high-traffic areas [19].

1.4 RECONFIGURATION OF IoT PROCESSORS

The right configuration of a processor for a specific application plays an important role. It is one of the major challenges for IoT processor designers to determine the best processor configurations that can meet the execution requirements of the applications. Table 1.2 presents the configuration of IoT processors in terms of the number of cores, CPU/GPU/DSP, on-chip/off-chip memory, power consumption, and pipeline.

In Table 1.2, we have presented examples of specific state-of-the-art processors to understand the configurations, though these configurations are for representative purpose only. The trade-off performance, area, and cost must be considered while selecting the processor for IoT application. Also, one must know the degree of reconfigurability or ability to adapt to the changes of configuration during the runtime by heterogeneous processors that provide a variety of processing resources for executing the applications.

TABLE 1.2
Sample Configuration of Processors Widely Used in IoT Applications

Parameters	Configurations			
	1	2	3	4
CPU/GPU	Arm Cortex-A7	Arm Cortex-A15	Microchip SAM R34/35	STM32H7
Frequency	500 MHz	1.9 GHz	2.4 GHz	550 MHz
Cores	2	4	1–4	2
Pipeline	4	15	4	–
Cache	32 KB L1, 1 MB L2	32 KB L1, 2 MB L2	256 KB Flash	16 KB L1, 16 KB L2
Memory	2 GB RAM	1 TB RAM	40 KB RAM	1 MB RAM
Execution	In-order	Out-of-order	Out-of-order	Out-of-order

In processors, many things can be configured such as instruction queue [20], buffer reorder [21], pipelines [22], and register files [23].

Among these, memory is one of the most important components that governs the performance and energy.

An IoT processor shall be equipped with many features like spatial and temporal locality of the IoT applications. The advanced reconfiguration of memory techniques can reduce power consumption up to 62% [24]. The cache memory parameters can be specified or changed during the run-time based on the application. There are some challenges in reconfigurability such as augmenting cache memory, algorithm tuning, and cache tuning.

1.5 CONCLUSION

IoT has proven to be one of the important needs for all sectors. The growth of IoT applications also demands scalable processor architectures. Because the technology is changing quickly, it is very difficult to meet the architectural requirements for these cases. IoT designers are able to measure both hardware and software to satisfy these design targets of processor products. In this chapter, we have presented a detailed discussion on the architecture along with the characteristics. Intel and ARM processors are the highly popular low-power processors. We have also presented a sample configuration of processors widely used in IoT applications.

REFERENCES

1. M. A. Khan, "An IoT Framework for Heart Disease Prediction Based on MDCNN Classifier," in *IEEE Access*, vol. 8, pp. 34717–34727, 2020, doi: 10.1109/ACCESS.2020.2974687.
2. M. A. Khan and K. A. Abuhasel, "Advanced Metameric Dimension Framework for Heterogeneous Industrial Internet of Things," in *Computational Intelligence*, vol. 37, pp. 1367–1387, 2021, https://doi.org/10.1111/coin.12378.

3. M. A. Khan and K. A. Abuhasel, "An Evolutionary Multi-hidden Markov Model for Intelligent Threat Sensing in Industrial Internet of Things," in *Journal of Supercomputing*, vol. 77, pp. 6236–6250, 2021, https://doi.org/10.1007/s11227-020-03513-6.
4. M. A. Khan and N. S. Alghamdi, "A Neutrosophic WPM-Based Machine Learning Model for Device Trust in Industrial Internet of Things," in *Journal of Ambient Intelligence and Humanized Computing*, 2021, https://doi.org/10.1007/s12652-021-03431-2.
5. N. S. Alghamdi and M. A. Khan, "Energy-Efficient and Blockchain-Enabled Model for Internet of Things (IoT) in Smart Cities," in *Computers, Materials & Continua*, vol. 66, no. 3, pp. 2509–2524, 2021.
6. Mahmoud Khalifa, Fahad Algarni, Mohammad Ayoub Khan, Azmat Ullah and Khalid Aloufi, "A Lightweight Cryptography (LWC) Framework to Secure Memory Heap in Internet of Things," in *Alexandria Engineering Journal*, vol. 60, no. 1, pp. 1489–1497, 2021, ISSN 1110-0168, https://doi.org/10.1016/j.aej.2020.11.003.
7. W. U. Khan, X. Li, A. Ihsan, M. A. Khan, V. G. Menon and M. Ahmed, "NOMA-Enabled Optimization Framework for Next-Generation Small-Cell IoV Networks Under Imperfect SIC Decoding," in *IEEE Transactions on Intelligent Transportation Systems*, doi: 10.1109/TITS.2021.3091402.
8. S. Nandy, M. Adhikari, M. A. Khan, V. G. Menon and S. Verma, "An Intrusion Detection Mechanism for Secured IoMT Framework Based on Swarm-Neural Network," in *IEEE Journal of Biomedical and Health Informatics*, doi: 10.1109/JBHI.2021.3101686.
9. A. Munusamy *et al.*, "Edge-Centric Secure Service Provisioning in IoT-Enabled Maritime Transportation Systems," in *IEEE Transactions on Intelligent Transportation Systems*, doi: 10.1109/TITS.2021.3102957.
10. S. Verma, S. Kaur, M. A. Khan and P. S. Sehdev, "Toward Green Communication in 6G-Enabled Massive Internet of Things," in *IEEE Internet of Things Journal*, vol. 8, no. 7, pp. 5408–5415, April 1, 2021, doi: 10.1109/JIOT.2020.3038804.
11. L. Xu, X. Zhou, M. A. Khan, X. Li, V. G. Menon and X. Yu, "Communication Quality Prediction for Internet of Vehicle (IoV) Networks: An Elman Approach," in *IEEE Transactions on Intelligent Transportation Systems*, doi: 10.1109/TITS.2021.3088862.
12. A. Munusamy *et al.*, "Service Deployment Strategy for Predictive Analysis of FinTech IoT Applications in Edge Networks," in *IEEE Internet of Things Journal*, doi: 10.1109/JIOT.2021.3078148.
13. A. Mukherjee, P. Goswami, M. A. Khan, L. Manman, L. Yang and P. Pillai, "Energy-Efficient Resource Allocation Strategy in Massive IoT for Industrial 6G Applications," in *IEEE Internet of Things Journal*, vol. 8, no. 7, pp. 5194–5201, April 1, 2021, doi: 10.1109/JIOT.2020.3035608.
14. https://www.intel.com/content/www/us/en/products/details/processors.html.
15. B. Joseph, "Scaling for IoT Market Demands," https://www.digit.in/features/apps/scaling-for-iot-market-demands-34645.html, published on March 15, 2017.
16. C. Perera, A. Zaslavsky, P. Christen and D. Georgakopoulos, "Context Aware Computing for the Internet of Things: A Survey," in *IEEE Communications Surveys & Tutorials*, vol. 16, no. 1, pp. 414–454, 2014.
17. K. Ashton, "That 'internet of things' thing," in *RFiD Journal*, vol. 22, no. 7, pp. 97–114, 2009.
18. V. S. Gopinath, J. Sprinkle and R. Lysecky, "Modeling of Data Adaptable Reconfigurable Embedded Systems," in *2011 18th IEEE International Conference and Workshops* on *Engineering of Computer Based Systems (ECBS)*, pp. 276–283, 2011.
19. M. A. Khan, M. T. Quasim, F. Algarni and A. Alharthi, "Internet of Things: On the Opportunities, Applications and Open Challenges in Saudi Arabia," in *2019 International Conference on Advances in the Emerging Computing Technologies (AECT)*, pp. 1–5, 2020, doi: 10.1109/AECT47998.2020.9194213.

20. D. Folegnani and A. González, "Energy-Effective Issue Logic," in *ACM SIGARCH Computer Architecture News*, vol. 29, pp. 230–239, 2001.
21. Y. Kora, K. Yamaguchi and H. Ando, "MLP-Aware Dynamic Instruction Window Resizing for Adaptively Exploiting Both ILP and MLP," in *Proceedings of the 46th Annual IEEE/ACM International Symposium on Microarchitecture*, pp. 37–48, 2013.
22. J. Abella and A. González, "On Reducing Register Pressure and Energy in Multiple-Banked Register Files," in *Proceedings of the 21st International Conference on Computer Design, 2003*, pp. 14–20, IEEE, 2003.
23. A. Efthymiou and J. D. Garside, "Adaptive Pipeline Structures for Speculation Control," in *Proceedings of the Ninth International Symposium on Asynchronous Circuits and Systems, 2003*, pp. 46–55, 2003.
24. A. Gordon-Ross, F. Vahid and N. D. Dutt, "Fast Configurable-Cache Tuning with a Unified Second-Level Cache," in *IEEE Transactions on Very Large-Scale Integration (VLSI) Systems*, vol. 17, no. 1, pp. 80–91, 2009.

2 Design Methods and Approaches for IoT Hardware

Ali Nezaratizadeh
Shahid Rajaee Teacher Training University
Tehran, Iran

CONTENTS

- 2.1 Introduction 14
  - 2.1.1 IoT Network Topologies 15
    - 2.1.1.1 Point to Point 15
    - 2.1.1.2 Bus 15
    - 2.1.1.3 Star 16
    - 2.1.1.4 Ring 17
    - 2.1.1.5 Mesh 17
    - 2.1.1.6 Hybrid 17
- 2.2 Architecture of IoT Network 18
  - 2.2.1 Mainboard 18
  - 2.2.2 Gateways 18
  - 2.2.3 Links 18
  - 2.2.4 Communication Protocols 19
    - 2.2.4.1 Wired-Modbus Protocol 19
    - 2.2.4.2 Wireless-IEEE 802.11 Standards 20
  - 2.2.5 Serial/Parallel Communication 20
  - 2.2.6 Physical Layer Standard 21
    - 2.2.6.1 TIA/EIA 232, RS-232 21
    - 2.2.6.2 TIA/EIA 422, RS-422 22
    - 2.2.6.3 TIA/EIA 485, RS-485 23
    - 2.2.6.4 Power Line Communication Using RS-485 24
    - 2.2.6.5 Wireless 25
  - 2.2.7 Recommendation for IoT Serial Communication 27
  - 2.2.8 Sensors and Actuators Nodes 28
    - 2.2.8.1 Sensors 28
    - 2.2.8.2 Actuators 28
  - 2.2.9 Summarizing IoT Block Diagram 30

2.3 A Scenario of Switching an IoT Light State 30
    2.3.1 The Initialization Step of Powering Up the Node and Connect
           It to the Mainboard 30
    2.3.2 Ending Data from the Sensor to the Mainboard 30
    2.3.3 Blue Color Blocks 31
    2.3.4 Step 1: Smart Switch Sends Data to the Mainboard 31
    2.3.5 Step 2: Mainboard Sends Data to the Actuator 31
    2.3.6 Step 3: Actuator Acts and Sends an Optional Report to the Mainboard .. 32
    2.3.7 Step 4 (Optional) 32
2.4 IoT Basic Circuit Design 32
    2.4.1 From Analog to Digital Signal 32
2.5 A Practical Example of an IoT Hardware 36
    2.5.1 Wired 36
    2.5.2 Wireless 36
2.6 Conclusion 36
References 37

2.1 INTRODUCTION

Using energy and material as efficiently as possible in today's world is a challenge. With the help of the Internet and computer networks, it is possible to dedicate an IP address to the equipment. For example, an energy monitoring system can calculate the amount of energy that the user consumes. By analyzing the output report of this system, the user can choose better time scheduling for turning ON and OFF the lighting system to curtail the use of energy.

In this chapter, after explaining conceptual Internet of Things (IoT) hardware design, each sub-part is analyzed in subsequent sections. Firstly, a block diagram of the whole system is explained. After that mainboard, sensors, and actuators are explained which are the sub-parts of the main block diagram. In both sensor and actuator sections, apart from the IoT application, some required basic knowledge of electronics is mentioned. Finally, by explaining a scenario of switching an IoT light state context is reviewed in a practical simple system.

An important question can be "What is the aim of using IoT device network dispite the extensive use of a powerful computer network architecture? In other words, what are the factors that lead to the new IoT network design?"

Firstly, for evaluating a network there are several criteria, which are called quality of service (QoS). This term provides predictable and consistent data transfer services while network resources are used as efficiently as possible. A computer network is something general but IoT devices are mostly optimized for specific domains. For instance, in an industry, reliability is of importance but for a simple smart lock, security is a non-negligible feature. Second, in IoT networks, another vital factor is the cost of the devices. IoT hardware can be designed a lot cheaper than computer network hardware. In one system engineers need to have redundancy, in another one the need of real-time communication is vital. Sometimes, the electronic required power and the costs would matter in mass production. As a result of these factors, developing a custom computer network for IoT devices is essential. Briefly, IoT

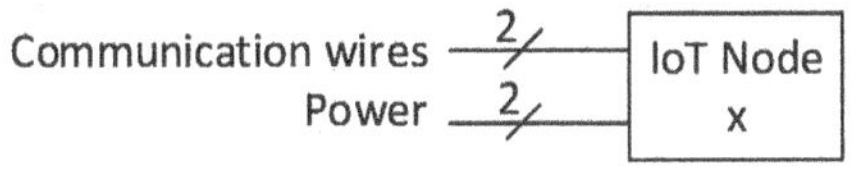

FIGURE 2.1 An IoT node *x* symbol and its output communication wire.

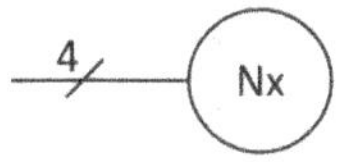

FIGURE 2.2 Simplified IoT node *x*.

communication is a simplified form of a computer network in which the engineers strengthen required parts according to the system requirements.

2.1.1 IoT Network Topologies

Network topologies define how physically IoT nodes connect. Figure 2.1 shows a schematic representation of an IoT device of number *x*. That is named as node *x*. This specific node needs four wires to connect to a network. Two wires dedicate to data communication and remaining are used as power supply wires.

In Figure 2.2 above node is simplified to a simpler representation that utilizes extensively in physical network topology diagrams.

2.1.1.1 Point to Point

Two nodes are connected by one dedicated link. An example is connecting the mobile phone to a laptop for transmitting data or screen sharing. Figure 2.3 shows the point-to-point network topology.

2.1.1.2 Bus

The message is transmitted and delivered to all nodes. Nodes need to process delivered messages and catch their own by matching addresses. This topology has some drawbacks namely single point of failure, all nodes process broadcasted data which increases overhead computing. Moreover, security can be an issue in this system because all nodes have access to other nodes' data. Figure 2.4 shows the bus topology.

FIGURE 2.3 Point-to-point network topology.

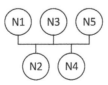

FIGURE 2.4 Bus topology.

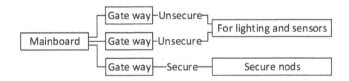

FIGURE 2.5 Proposed bus topology for building management system for secure and unsecure nodes.

This topology is extensively used in building management systems (BMS) because of the cost and simplicity of implementation. The mentioned drawbacks of bus topology can be accepted in BMS as specific application. Firstly, the size of data is relatively small, as a result, the processing overhead is negligible take switching lamp state as an example. The second challenge is a single point of failure, which can be alleviated by using a second redundancy network. And concerning the security, using a dedicated secure transmission medium is suggested, namely burglar systems or smart locks. Figure 2.5 shows the proposed bus topology for BMS secure and unsecure nodes.

2.1.1.3 Star

In this topology, all nodes are connected to the central hub. This topology like the bus topology is a single point of failure because of the central hub. This type is mostly applied to IoT wireless nodes. An access point works as a central hub that connects all wireless nodes to the mainboard as controller. Figure 2.6 shows star topology in which N1–N4 can have communication through N3.

In the case of using wired nodes, the cost of the network implementation in the star topology is much more than the bus topology. In Figure 2.7, both types of topologies are compared. Each node, connect to the central node (mainboard) via Gateway (GW). In wireless type because of the coverage area of antenna, it is possible to reduce the number of network equipment.

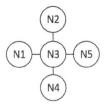

FIGURE 2.6 Star topology.

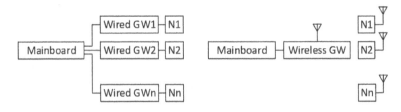

FIGURE 2.7 Comparison of star topology for wired and wireless BMS network.

Design Methods and Approaches for IoT Hardware 17

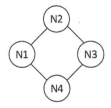

FIGURE 2.8 Ring topology.

2.1.1.4 Ring
In this topology, data travel from one node to another node until it reaches the destination (N4). This topology has several advantages such as one direction of data flow leads to a reduction of data collusion; The problem with this network is that if one of the nodes loses its connection to the rest of the network, the performance of the whole network will be disrupted. Figure 2.8 shows the ring topology.

2.1.1.5 Mesh
Each node is connected to another node and the cost of implementation is high. This topology is mostly used for wireless networks. Figure 2.9 shows the mesh topology.

2.1.1.6 Hybrid
In this topology, two or more different networks are used. For example, in a smart home network, one might prefer to use wireless and wired nodes together. As a result, it is possible to cover the weaknesses of a network with other network features. In Figure 2.10, N1–N4 are wireless nodes and N5 is wired node. Figure 2.11 also presents hybrid wireless and wired network block diagram.

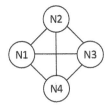

FIGURE 2.9 Mesh topology.

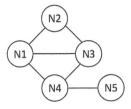

FIGURE 2.10 Hybrid topology.

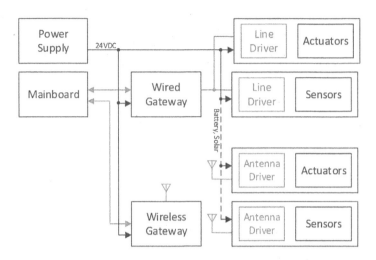

FIGURE 2.11 IoT network block diagram.

2.2 ARCHITECTURE OF IoT NETWORK

In general, IoT hardware is divided into four parts—mainboard, gateways, nodes, and links. Firstly, Firstly, each part will be described, and finally, using these parts, a simple scenario of turning a lamp on and off will be described.

2.2.1 MAINBOARD

The mainboard is responsible for managing this network, it is also named controller. All sensors and actuators are connected to the mainboard via gateways. Mainboard obtains sensors' data then by considering pre-defined rules, mainboard decides to activate a specific actuator. This mainboard also provides a graphic user interface that helps the user to control the system manually, define rules, and scenarios.

Mainboard hardware is usually a small single-board computer like Raspberry Pi. Figure 2.12 shows Raspberry Pi 4. This small computer has enough capabilities for managing an IoT nodes of a BMS. Its price starts from 35$. USB ports are mostly used for connecting the gateways to the Pi, and by a Wi-Fi hotspot, a mobile phone can be used to connect to the BMS controller web app.

2.2.2 GATEWAYS

Gateways are the interface that converts different mediums to standard understandable mediums for the mainboard. For example, there are ten IoT nodes in one IoT network and five of them are wired and remaining are wireless type. The gateway must standardize these two to a USB standard port.

2.2.3 LINKS

Links connect nodes, gateways, and mainboards for communication such as a wire.

Design Methods and Approaches for IoT Hardware

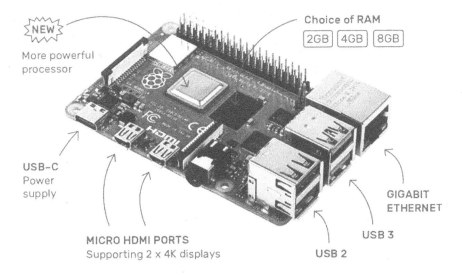

FIGURE 2.12 Raspberry Pi 4.

2.2.4 Communication Protocols

A communication protocol is a set of rules that allow nodes to communicate via links. The development of widely used IEEE 802.11 and 802.3 ("Wireless LAN Medium Access Control [MAC] and Physical Layer [PHY] Specifications" 2018; "IEEE Standard for Ethernet" 2020) standards makes it a proper choice for wireless and wired IoT nodes. Many companies now use their non-standard closed protocol to meet their network topology needs. In this chapter, Modbus standard protocol is described.

2.2.4.1 Wired-Modbus Protocol

Modbus was firstly used by Modicon in 1996 for linking programmable logic controllers. After several years, Modbus has become an open protocol and several industrial and home automation systems are based on Modbus protocol. There are two types of Modbus transmission modes: ASCII and RTU. The ASCII mode uses ASCII characters for creating a transmission data frame. Table 2.1 details Modbus ASCII message frame (Modicon 1996), which contains necessary data for the master to connect to nodes. It contains the address of the node, function, and the data. The

TABLE 2.1
Modbus ASCII Message Frame

Start	Address	Function	Data	LRC Check	End
					2 Chara
1 Char	2 Chars	2 Chars	n Chars	2 Chars	CRLF

Step 1: M1 sends data to S1

Step 2: S1 sends back massage to M1 to confirm message received

FIGURE 2.13 Modbus protocol sending data from master (M1) to slave (S2).

RTU mode uses binary coding and there is CRC error checking. This protocol uses RS-232, RS-485, or RS-422 as a physical layer. RS-232 can connect only two nodes. For more than two nodes, RS-485 and RS-422 can make proper connection.

Figure 2.13 shows how data transmit from master node M1 to slave S2. Firstly, the master generates a message frame. After that, this message is broadcasted to all slaves. Only one of them, which matches to the address of the message frame, starts to compute the message although all other nodes receive the same message. Next, if the message is received successfully, message is sent back from the slave to the master to confirm the successful receipt.

2.2.4.2 Wireless-IEEE 802.11 Standards

IEEE 802 standards are available in the IEEE GET 802™ program ("GET 802(R) Standards" n.d.). These standards have been published in PDF for six months. Table 2.2 summarizes the standards number and their titles. Many low-cost Wi-Fi modules like ESP32 are based on IEEE 802.11. This module has many capabilities that make it a proper choice for IoT wireless nodes. ESP32 is a single 2.4 GHz Wi-Fi-and-Bluetooth combo chip designed with the TSMC ultra-low-power 40-nm technology ("ESP32 Series Datasheet" 2021).

2.2.5 Serial/Parallel Communication

After creating a data frame, for sending data from the transmitter to receiver, there are two possibilities: parallel or serial communications. In a parallel communication, 8 bits are sent simultaneously in one clock by 8 wires. Sending massive data

TABLE 2.2
IEEE 802 Standards

IEEE Standard	Title
IEEE 802(R)	Overview and Architecture
IEEE 802.1	Bridging and Management
IEEE 802.3	Ethernet
IEEE 802.11	Wireless LANs
IEEE 802.15	Wireless PANs
IEEE 802.16	Broadband Wireless MANs
IEEE 802.19	TV White Space Coexistence Methods
IEEE 802.21	Media Independent Handover Services
IEEE 802.22	Wireless Regional Area Networks

Design Methods and Approaches for IoT Hardware 21

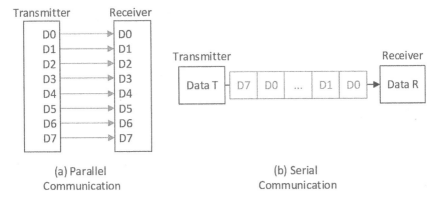

FIGURE 2.14 (a) Parallel communication and (b) serial communication.

in a short time is a characteristic of parallel communication, for example, reading data from DDR3 RAM. But in a serial communication, 1 bit is sent in one clock by just one I/O pin. The data rate is also less than the parallel type but sufficient for IoT communication. And the cost of copper wire and I/O pins are high. As a result, serial communication is the proper choice for wired IoT devices. Figure 2.14 shows both serial and parallel communications.

2.2.6 Physical Layer Standard

According to the OSI (Open Systems Interconnection) model, a physical layer has different components. But in this section, only the medium is described. As mentioned in previously for IoT devices, serial communication is preferred. For serial communication, different standards are used. Each standard and its networking requirements determine the number of communication wires. For example, for sending differential serial data, each IoT node needs two twisted-pair copper wires as a transmission which is driven by line driver ICs.

In the following context, RS-232, RS-422, and RS-485 standards are described. In contradiction of complete interface standards, which define functional data frame specifications, these are electrical-only standards. Between these three, only RS-485 standards can satisfy the needs of IoT daisy-chain or bus network. Moreover, using power lines to transmit the data is explained and the wireless type is briefly described. Finally, the recommendations for IoT serial communication are suggested.

2.2.6.1 TIA/EIA 232, RS-232

The CMOS TTL voltages are typically in the 3.3–5 V range, while the RS-232 can be 12 V. To meet the voltage requirement of RS-232, it is possible to convert +5 V TTL available voltage to ±12 V by circuit named "dc-dc converter," which has different types. MAX-232 IC converts TTL to RS-232 data in which there is an integrated charge pump boost converter to increase +5 V input voltage to RS-232 required level. Figure 2.15(a) shows that TTL data amplitude ranges from 0 V to +5 V, which is converted to RS-232 standard at the output. Figure 2.15(b) shows two blue probs

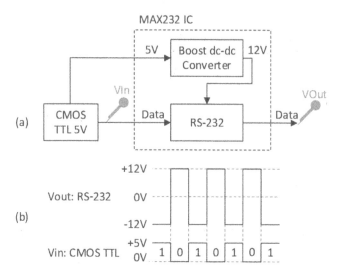

FIGURE 2.15 TTL to RS-232 (a) block diagram and (b) waveform.

FIGURE 2.16 Single-ended, unidirectional, half duplex.

voltage waveforms. For each logic 1 TTL level (+5 V) as input, output goes to −12 V, and for logic 0 TTL level (0 V), the output goes to +12 V.

As illustrated in Figure 2.16, the data go in one direction. For implementing full-duplex communication (sending and receiving), this block diagram should be repeated in the inverse direction. Portable and handheld apparatus use RS-232 standard. This standard is suitable for low data rates, low range, and short-run applications. If one needs a higher data rate, longer distance, and more than one receiver node, RS-422 could be the appropriate choice.

2.2.6.2 TIA/EIA 422, RS-422

RS-422 is referred as a balanced differential signaling standard. In this type, data are transmitted by two twisted-pair copper wires. Because of the differential signaling, common-mode noise, which is induced to twisted pair copper wire, will be canceled. Data line can be as long as 4,000 feet with a data rate of around 100 kbps. And for short distance, the data transfer speed can be up to 10 Mbps.

Figure 2.17 illustrates RS-422 half-duplex differential signaling. In this diagram, TTL digital data are converted into differential signals and then feed to the transmission line. In the receiver, these two differential signals convert to TLL data again. This standard can provide service up to ten receivers. However, this type is half-duplex for overcoming all limitations, thus RS-485 is introduced in the next section.

Design Methods and Approaches for IoT Hardware 23

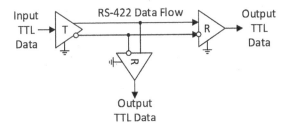

FIGURE 2.17 RS-422 half-duplex differential signaling.

2.2.6.3 TIA/EIA 485, RS-485

In this standard, serialized input data change to differential pair A and B, which is similar to RS-422. Figure 2.18 shows the ADM485 block diagram in which enable pins (RE and DE) are key elements to have a bidirectional network. A and B pins are three-stated gates controlled by RE and DE pins. These pins' functionality can be changed to input, output, or high impedance. By enable pins, the working modes of the chip can be set to receiver or driver (transmitter). Thus, by MCU software programming, the direction of the data flow can easily change. In most applications, RE and DE pins are tied together as RE is an "active-low enable pin" and DE is inverse. In RS-485, only one driver can be active on the bus at any time and other nodes are in receiver mode. This process must be controlled through software to avoid any data collision.

Here, Table 2.3 summarizes ADM485 pin function descriptions.

For sending "Logic 1" A = 1, B = 0 and for "Logic 0" A = 0 and B = 1. Inversely in the receiver, by differentiating A and B pins voltage, serial data can be generated. According to Table 2.4, if the receiver VIA − VIB ≥ 200 mV output serial is "1" and if VIA − VIB ≤ 200 mV the output is "0."

In the industry, TIA/EIA-485-A transmission line standard is widely used such as Profibus and Modbus (Marais 2008; "Interface Circuits for TIA/EIA-485 (RS-485)" 2007). Differential signaling rejects common-mode noise, which makes it suitable

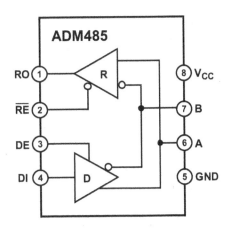

FIGURE 2.18 ADM485 block diagram.

TABLE 2.3
ADM485 Pin Function Descriptions

Pin No.	Mnemonic	Function
1	RO	Receiver output. When enabled, if A is greater than B by 200 mV, RO is high. If A is less than B by 200 mV, RO is low.
2	RE	Receiver output enable. A low level enables the receiver output, RO. A high level places it in a *high impedance state*.
3	DE	Driver output enable. A high level enables the driver differential outputs, A and B. A low level places it in a *high impedance state*.
4	DI	Driver input. When the driver is enabled, a logic low on DI forces A low and B high, while a logic high on DI forces A high and B low.
5	GND	Ground connection, 0 V.
6	A	Noninverting receiver input A/driver output A.
7	B	Inverting receiver input B/driver output B.
8	VCC	Power supply, 5 V ± 5%.

Source: "ADM485 Datasheet | Analog Devices" (n.d.).

for long-distance communication (~1 km) (2011). ADM485 can only drive 32 nodes and for more, it needs a repeater. Figure 2.19 shows cable length versus data rate for RS-485. As cable length is increased, a lower data rate can be transmitted.

2.2.6.4 Power Line Communication Using RS-485

IoT nodes need power supply to power up, which adds two additional wires to the RS-485 data wire. Thus, each node needs four wires to work properly. By considering the power supply as DC signal and RS-485 data as AC signal, it is feasible to reduce the number of wires to just two by feeding both AC and DC signal in two wires. "Power Line Communication Using RS-485 Simulation Reference Design" (2018) establishes a simulation model for implementing RS-485 communication over power cabling.

TABLE 2.4
Differential Receiver Truth Table

RE	A − B (Inputs)	RO
0	≥+200 mV	1
0	≤−200 mV	0
0	−200 mV ≤ (A − B) ≤ +200 mV	X
1	X	High-Z

Design Methods and Approaches for IoT Hardware

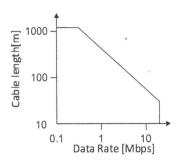

FIGURE 2.19 Cable length versus data rate for RS-485. (From Kugelstadt 2008.)

2.2.6.5 Wireless

For wireless communication, engineers, by using Maxwell equation, send data to free space impedance $\eta_0 = \sqrt{\mu_0/\epsilon_0} = 733\,\Omega$ (Pozar 2011) via an antenna. The digital data are firstly modulated to high-frequency signals. This frequency can be as high as Wi-Fi 2.4 GHz or as low as AM radiofrequency. As the frequency increases, the antenna size decreases, which is important in portable devices. ASK RF modules accept digital data at the input terminal of the transmitter and in the receiver module, RF data after demodulation is converted into digital data. Figure 2.20 shows the ASK module block diagram. Figure 2.21 shows the signal waveform of ASK modulating. As it is illustrated, carrier frequency and digital data are multiplied with each other to generate an output ASK signal.

ASK modules in the market have two different carrier frequencies: 433 and 315 MHz. These modules only send digital data and do not take into account whether the data are received correctly. As a result, software should do many other tasks such as adding CRC to the digital data frame or check if space is not occupied by another RF transmitter to send data. These modules mostly are used for simple two-node communication such as car RF remote controller. Figure 2.22 shows the transmitter and receiver of Chinese ASK modules.

For more complex networks, designers prefer to use Wi-Fi modules that are based on IEEE 802.11 standard. Moreover, other equipment such as Wi-Fi access points are already available in the market at a reasonable price. For example, Espressif Systems is one of the prominent companies in this area. ESP32 is a low-cost, low-power system on a chip microcontroller with integrated Wi-Fi and dual-mode Bluetooth. Thus, this module is capable of running a software algorithm, changing the status of output I/O pins, and also sending data wirelessly. This module has a different board version. One of them uses a patch (microstrip) antenna to decrease

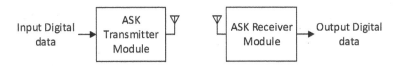

FIGURE 2.20 ASK module block diagram.

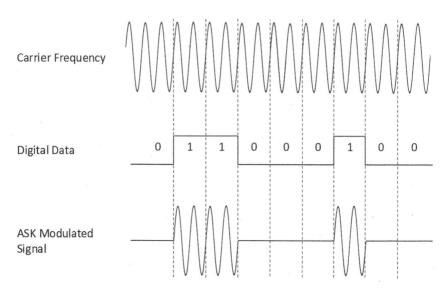

FIGURE 2.21 ASK modulation.

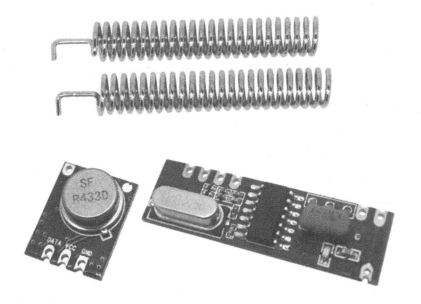

FIGURE 2.22 Chinese ASK modules.

the size of the PCB. Many approaches are used to miniaturize planar-type antennas such as microstrip (Balanis 2016). Figure 2.23 shows ESP32 Wi-Fi module with its golden 2.4 GHz microstrip antenna and U. FL connector for external antenna. All in all, one can use this module for IoT devices without thinking about wireless communication, data loss, node collision, and many other complex RF requirements in designing RF wireless nodes from scratch.

Design Methods and Approaches for IoT Hardware

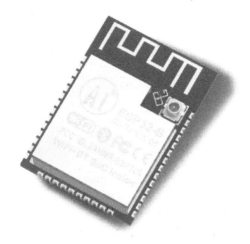

FIGURE 2.23 ESP32 and microstrip antenna.

2.2.7 Recommendation for IoT Serial Communication

RS-232, RS-422, and RS-485 are three famous serial communication electrical standards, which have different features and limitations such as signaling techniques, communication modes, and network complexity. For low-cost, long-run, reliable, low-data rate communication and networking, RS-232 and RS-485 are preferred to use. Table 2.5 compares three serial communication standards. RS-485 daisy-chain network is preferred to use in an IoT wired network because of higher distance, number of nodes, and data rate (Marais 2008; "Interface Circuits for TIA/EIA-485 (RS-485)" 2007; Hazen 2003; "AN-914 Understanding Power Requirements in RS-232 Applications" 2013).

For long-run IoT devices, wired communication is preferred. Since wireless node quality depends on many criteria such as weather. In a rainy environment, RF signal

TABLE 2.5
RS-232, RS-422, and RS-485 Comparison

RS-232	RS-232	RS-422	RS-485
Signaling Technique	Single-Ended (unbalanced)	Differential (balanced)	Differential (balanced)
Drivers and Receivers on Bus	1 Driver 1 Receiver	1 Driver 10 Receivers	32 Drivers 32 Receivers
Maximum Cable Length	50 feet	4000 feet	4000 feet
Original Standard Maximum Data Rate	20 kbps	10 Mbps down to 100 kbps	10 Mbps down to 100 kbps

Source: Hazen (2003).

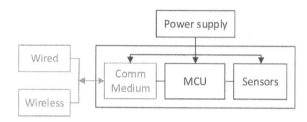

FIGURE 2.24 Sensors block diagram.

attenuation is high and nodes might lose their connections. Moreover, if the security of the system is important, someone can more easily access the wireless node than a local wired node. Another point is the aging of electronic components that has much more effect on transmitting power of RF node than wired node.

2.2.8 Sensors and Actuators Nodes

Figures 2.24 and 2.25 show sensors and actuators block diagrams. These two are briefly explained as follows.

2.2.8.1 Sensors
Sensor nodes can send measurement data to the mainboard by a communication link. This part firstly describes the definition of sensors and their tasks; after that, a story starts from an analogue simple sensor and finishes to a complex digital one.

A sensor is a device that reacts to physical or chemical actions or states. For example, light-dependent resistor (LDR) sensors can change their electrical properties by exposing their surface to the light.

2.2.8.2 Actuators
Mainboard sends data to actuator nodes for performing specific tasks such as turn the light ON or OFF. In smart homes and industry, actuators can be a mechanical relay or contactor, or solid-state relay. These two devices can switch power lines to drive even a high-power motor. In general, inside each node, there is an MCU. The

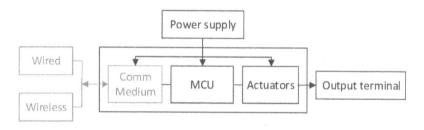

FIGURE 2.25 Actuators block diagram.

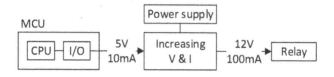

FIGURE 2.26 Driving mechanical relay by the microcontroller block diagram.

output pins are not capable of providing high current to drive a relay. As a result, a small circuit that accepts low current low voltage at the input and provides high voltage high current at the output pin is required. Figure 2.26 shows a block diagram of driving a mechanical relay using an MCU. "STM32 GPIO Configuration for Hardware Settings and Low-Power Consumption" (2017) describes ST 32-bit microcontroller general-purpose for input/output pins. The maximum output current by any I/O is 25 mA and the sum of all I/O current is 80 mA.

One can use a simple common emitter transistor and fast diode for driving a mechanical relay. Relay driver integrated circuit (IC) such as ULN2803 also performs the task well. The ULN2803A device is a 50 V, 500 mA Darlington transistor array. The device consists of eight NPN Darlington pairs that feature high-voltage outputs with common-cathode clamp diodes for switching inductive loads ("ULN2803A Darlington Transistor Arrays" 2017). Figure 2.27 shows the ULN2803

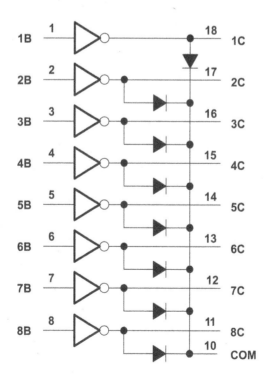

FIGURE 2.27 ULN2803A logic diagram.

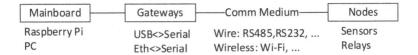

FIGURE 2.28 Summarizing IoT block diagram.

logic diagram. 1B to 8B are input and 1C to 8C are high voltage and current outputs, which can connect to relay to turn it ON and OFF.

2.2.9 Summarizing IoT Block Diagram

Briefly, an IoT network consists of four parts: mainboard, gateways, nodes, and links. Mainboard is the most complex part and is preferred to be bought from the market such as Raspberry Pi or using fanless mini-PC. Gateways are the interface that connects serial nodes to available ports (USB, Ethernet) of Raspberry Pi (Figure 2.28).

2.3 A SCENARIO OF SWITCHING AN IoT LIGHT STATE

For simplicity, a scenario of switching an IoT light state is described, which can be generalized to a more complex system. In this scenario, it is assumed that the data are transmitted and received by nodes without any loss. Figure 2.30 shows the block diagram of switching an IoT light state ON or OFF. It contains four steps. The dashed sections are optional and can be varied in different architectures. The blue blocks of all steps are similar to each other which is related to wired communication links. Figure 2.29 shows the bus network topology of switching an IoT light state. In this figure, the pushbutton and actuator node connect to the gateway via wire and communicate to each other by the mainboard.

2.3.1 The Initialization Step of Powering Up the Node and Connect It to the Mainboard

After powering up, an IoT node tries to connect to the mainboard with its previous node address that is stored in its EEPROM. If this node address is already occupied in the address table of the mainboard (simple DHCP of the ethernet), the mainboard sends another free address to the node and the node starts to send data by the new dedicated address.

2.3.2 Ending Data from the Sensor to the Mainboard

Nodes can have different types of communication to the mainboard. For example, one node can send data after a specific time interval. Another node sends its state just

FIGURE 2.29 Bus network topology of switching an IoT light state.

Design Methods and Approaches for IoT Hardware

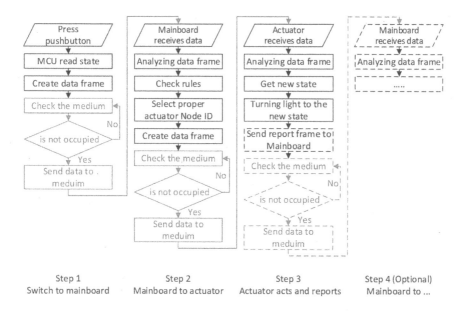

FIGURE 2.30 The block diagram of switching an IoT light state.

after an event is fired for instance, when the PIR sensor node detects the human IR. Sending real-time data in both wired and wireless networks occupies the transmission medium, that is why, data should be sent in a specific time interval. Since thermal conduction in the environment is slow, the temperature sensor needs to monitor the temperature every 5 minutes. However, the water valve of the smart garden needs a more high-speed communication link.

2.3.3 Blue Color Blocks

The blue color blocks in Figure 2.30 take care of data collision in the medium. Before sending data over the line, they check that the line is not occupied by other nodes or mainboard. After initializing and presenting the node successfully to the mainboard, step 1 starts.

2.3.4 Step 1: Smart Switch Sends Data to the Mainboard

Firstly, IoT smart switch node pushbutton is pressed. Its state is read by an MCU pin and a proper data frame is created to send this event to the mainboard. 0 checks and sends data by wired medium to the mainboard.

2.3.5 Step 2: Mainboard Sends Data to the Actuator

In the mainboard after receiving the data frame completely and correctly, data frame chunks are processed and the receiver node ID and its message are extracted. Then, the mainboard checks pre-defined rules according to extracted data. These rules relate the switch node state to its actuator node. After that, mainboard makes another data frame to send to the actuator node, which is performed by blue color blocks.

2.3.6 STEP 3: ACTUATOR ACTS AND SENDS AN OPTIONAL REPORT TO THE MAINBOARD

After the mainboard sends the data to the actuator successfully, the actuator analyzes the data and extracts its state. Then, it sets the light state according to mainboard message.

2.3.7 STEP 4 (OPTIONAL)

This step is optional; in some systems, it might require making a closed-loop control system that reports to the mainboard whether the light turns on successfully or not.

2.4 IoT BASIC CIRCUIT DESIGN

2.4.1 FROM ANALOG TO DIGITAL SIGNAL

Analog sensors, as their name suggests, have an analog output. A well-known example of an analog sensor is LDR. Figure 2.31 shows the real sensor that has two output pins.

The resistor of these two terminals is varying by changing the light intensity. Figure 2.32 chart illustrates LDE resistance vs. light intensity. In a dark area, resistors are intended to increase sharply.

Most of the electronic integrated circuits need voltage or current at their input terminals to operate. Thus, a converter that accepts a resistor at its input and generates acceptable voltage output is required. Figure 2.33 is the whole process of converting light intensity to voltage. The second block can be as simple as a voltage divider. However, Figure 2.34 shows a more precise measurement of LDR, and (2.1) shows the transform function.

FIGURE 2.31 Light-dependent resistor (LDR) or photo resistor. (From "Simple Ambient Light Sensor Circuit I Analog Devices" n.d.)

Design Methods and Approaches for IoT Hardware 33

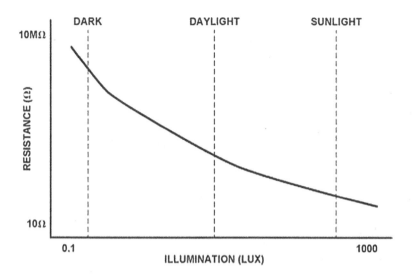

FIGURE 2.32 Sensor resistance vs. light intensity.

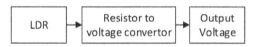

FIGURE 2.33 Light intensity to a voltage block diagram.

$$LDR = \frac{49.4\text{k}\Omega}{\frac{V_{OUT}}{V_{IN^+}} - 1} \quad (2.1)$$

The next sensor is more complex; this one is an IC. In an IC many features can be added that just a simple LDR can't have. For example, a closed-loop circuit that has

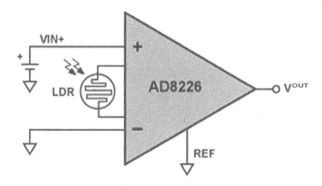

FIGURE 2.34 Simple circuit measures light intensity. (From "Simple Ambient Light Sensor Circuit | Analog Devices" n.d.)

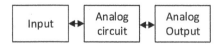

FIGURE 2.35 LM35 functional block diagram.

stable functionality in different situations or even a filter. Take NTC, for example, as a temperature-dependent resistor (Thermistor); for reading out this simple register as mentioned previously, we need to convert resistance to voltage to be acceptable for the next stage of the circuit. In this regard, an operation amplifier can do this task perfectly. It is possible to integrate the sensor and converter circuit into a single chip. This chip, apart from having a smaller size, consumes less power and is more affordable.

LM35 is an analog integrated temperature sensor. This sensor can easily measure environmental temperature with just three external components. The analog output of the sensor varies 10 mV per Celsius degree. Figure 2.35 shows LM35 IC inner circuit.

Up to here, two environmental parameters, light intensity and temperature change to an analog voltage. This analog signal needs to be converted into digital signals by analog to digital converter (ADC) so that it is acceptable for the IoT digital circuits. As a result, this part can also add another block to Figure 2.36 as an analog temperature sensor and convert it into a mixed-signal integrated circuit below the block diagram clarifying new digital temperature sensors. Figure 2.37 shows a block diagram of a digital integrated circuit temperature sensor.

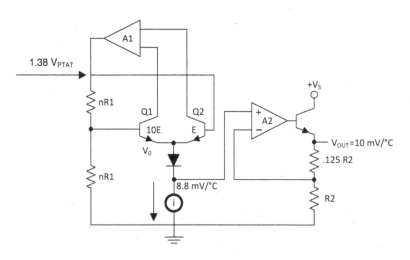

FIGURE 2.36 Analog sensors block diagram.

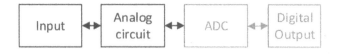

FIGURE 2.37 Digital temperature sensor block diagram.

Design Methods and Approaches for IoT Hardware 35

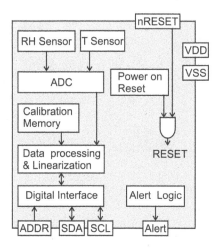

FIGURE 2.38 Functional block diagram of the SHT3x-DIS. The sensor signals for humidity and temperature are factory calibrated, linearized, and compensated for temperature and supply voltage dependencies. (From "SHT3x (RH/T)—Digital Humidity Sensor I Sensirion" n.d.)

Figure 2.38 SHT3x-DIS, a mixed-signal integrated circuit made by Sensirion company. As it is clear in the block diagram besides the ADC part, many other blocks enable this sensor to measure temperature to be intended to have 0.1°C precision for temperature and 1.5% RH for humidity.

Table 2.6 summarizes the pin assignment of SHT3x-DIS. Pins 1 and 2 accept digital data for communication. Figure 2.39 shows the sensor transparent top view. One might use an internal ADC microcontroller. The results can be acceptable if high precision does not matter. If the voltage or current temperature signal is transmitted

TABLE 2.6
SHT3x-DIS Pin Assignment

Pin	Name	Comments
1	SDA	Serial data; input/output
2	ADDR	Address pin; input; connect to either the logic high or low, do not leave floating
3	ALERT	Indicates alarm condition; output; must be left floating if unused
4	SCL	Serial clock; input/output
5	VDD	Supply voltage; input
6	nRESET	Reset pin active low; input; if not used it is recommended to be left floating; can be connected to VDD with a series a resistor of R ≥2 Kω
7	R	No electrical function; to be connected to VSS
8	VSS	Ground

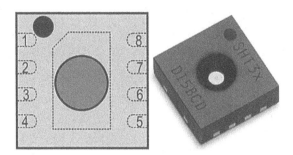

FIGURE 2.39 Sensor (transparent top view). Dashed lines are only visible if viewed from below. The die pad is internally connected to VSS.

via a copper line, analog signals will have noise, but if the output of the sensor is digitalized by its inner ADC, a digital signal is less prone to get noise.

2.5 A PRACTICAL EXAMPLE OF AN IoT HARDWARE

In this section, two types of IoT network are discussed. In both types, instead of designing boards from scratch, available modules in the market are utilized. For programming, the board Arduino is highly recommended.

2.5.1 Wired

For wired communication, three Arduino Nano boards are connected to each other by RS-485 converter modules. With the use of software, one of them is chosen to be the master and two others are made slaves. It is also recommended to add one LED as light to one slave and the other one plays smart switch node and implements a scenario of switching an IoT light state. Figure 2.40 shows a simple circuit diagram of connecting three Arduino Nano boards by RS-485 modules.

2.5.2 Wireless

For wireless IoT network, without a doubt using ESP32 or ESP8266 is the best choice. These modules can be programmed by Arduino. Moreover, many people share their open-source libraries to drive ICs or sensors. It is also noted that because of the I/O pins that this module has, one can use it as a wired node as well and connect RS-485 modules to them and make a hybrid topology.

2.6 CONCLUSION

In this chapter, IoT hardware design is described. Each topology is analyzed and their drawbacks are specifically investigated regarding IoT networks. A sub-part of the IoT network topology including nodes, links, gateways, and mainboard is also described. In general, these explanations help to choose proper topology and devices for IoT networks. For a wired IoT network, RS-485 standard and bus topology is

Design Methods and Approaches for IoT Hardware

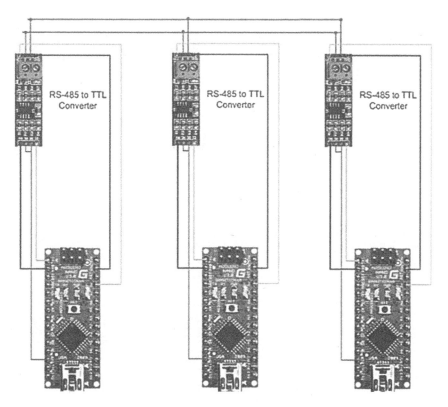

FIGURE 2.40 Connecting three Arduino Nano boards by RS-485 modules.

preferred. And for a wireless network, Wi-Fi standard is suggested because of the massive number of low-cost market equipment. One can use a laptop as a mainboard and ESP32 as a wired or wireless IoT node and evaluate different communication links that this device provides.

REFERENCES

"ADM485 Datasheet | Analog Devices." n.d. Accessed March 8, 2021. https://www.analog.com/en/products/adm485.html.

"AN-914 Understanding Power Requirements in RS-232 Applications." 2013. *Texas Instruments.* https://www.ti.com/lit/an/snla037b/snla037b.pdf

Balanis, Constantine A. 2016. *Antenna Theory: Analysis and Design.* John Wiley & Sons.

"ESP32 Series Datasheet." 2021. *Espressif Systems.* https://www.espressif.com/sites/default/files/documentation/esp32_datasheet_en.pdf

"GET 802(R) Standards." n.d. Accessed March 8, 2021. https://ieeexplore.ieee.org/browse/standards/get-program/page/series?id=68.

Hazen, Mark E. 2003. "Understanding Some Basic Recommended Standards for Serial Data Communications – A Comparison of RS-232, RS-422 and RS-485," 6. http://www.intersil.com/data/wp/WP0585.pdf.

"IEEE Standard for Ethernet." 2020. *IEEE*, June, 1–207. https://doi.org/10.1109/IEEESTD.2020.9146430.

"Interface Circuits for TIA/EIA-485 (RS-485)." 2007, 20. https://www.ti.com/lit/an/slla036d/slla036d.pdf

Kugelstadt, Thomas. 2008. "The RS-485 Design Guide: A Short Compendium for Robust Data Transmission Design." *Texas Instruments: Dallas, TX, USA*.

Marais, Hein. 2008. "RS-485/RS-422 Circuit Implementation Guide." *AN-960 Analog Devices*. https://www.analog.com/media/en/technical-documentation/application-notes/AN-960.pdf?doc=an-1177.pdf.

Modicon, I. 1996. "Modicon Modbus Protocol Reference Guide." *North Andover, Massachusetts*, 28–29.

"Power Line Communication Using RS-485 Simulation Reference Design." 2018, 26. https://www.ti.com/lit/ug/tiduei9/tiduei9.pdf

Pozar, David M. 2011. *Microwave Engineering*. John Wiley & Sons.

"SHT3x (RH/T) - Digital Humidity Sensor | Sensirion." n.d. Accessed February 20, 2021. https://www.sensirion.com/kr/environmental-sensors/humidity-sensors/digital-humidity-sensors-for-various-applications/.

"Simple Ambient Light Sensor Circuit | Analog Devices." n.d. Accessed February 17, 2021. https://www.analog.com/en/analog-dialogue/articles/simple-ambient-light-sensor-circuit.html.

"STM32 GPIO Configuration for Hardware Settings and Low-Power Consumption." 2017. *STMicroelectronics*, September. https://www.st.com/resource/en/application_note/an4899-stm32-microcontroller-gpio-configuration-for-hardware-settings-and-lowpower-consumption-stmicroelectronics.pdf

"ULN2803A Darlington Transistor Arrays." 2017, February. https://www.ti.com/lit/ds/symlink/uln2803a.pdf

"Wireless LAN Medium Access Control (MAC) and Physical Layer (PHY) Specifications." 2018. *IEEE*, August, 1–69. https://doi.org/10.1109/IEEESTD.2018.8457463.

3 IoT Solution Reference Architectures

Vinay Chowdary, Tiyasa Bera, and Arpit Jain
University of Petroleum and Energy Studies
Dehradun, India

CONTENTS

3.1 Introduction 39
3.2 Literature Review 40
3.3 Internet of Things (IoT) Reference Architecture 43
    3.3.1 Prerequisites for Reference Architecture 44
        3.3.1.1 Connection and Communications 44
        3.3.1.2 Gadget Management 44
        3.3.1.3 Security 45
3.4 The Layers of the IoT Architecture 45
    3.4.1 Three-Layer and Five-Layer Architectures 45
    3.4.2 Seven-Layer IoT Architecture 46
3.5 IoT Architecture Based on Fog Computing Paradigm 47
3.6 Edge Computing IoT Architecture 48
3.7 Hybrid IoT Architecture 48
    3.7.1 Edge Computing Layer 49
    3.7.2 Fog Computing Layer 49
    3.7.3 Cloud Computing Layer 49
3.8 Social IoT 49
    3.8.1 Basic Components 50
3.9 Conclusion 50
References 50

3.1 INTRODUCTION

The modern age is the age of science where we reached the zenith of civilization, and the Internet of Things (IoT) is the most important gift of science. Nowadays, we can control the whole world with the help of this IoT technology. IoT makes the bridge between the real world and the virtual world. Some basic components which are used in this technology are sensors for sensing the needs or commands which are to be performed, actuators are the devices that receive electrical input and give physical action, cloud for storing the data, and the Internet connection for controlling the whole process. Every day we are getting one step closer to the advanced universe and IoT is the road to it.

DOI: 10.1201/9781003122357-4

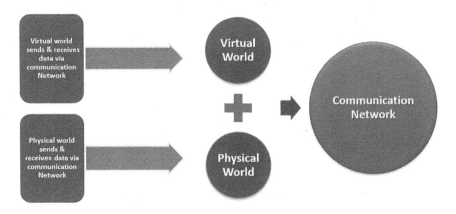

FIGURE 3.1 Internet of things block diagram.

IoT reference model and architecture or ARM mainly consists of two parts:

- Reference model: this model describes the domain using several sub-models. These are as follows:
  a. IoT Domain Model
  b. IoT Information Model
  c. IoT Functional Model
     i. IoT Communication Model
     ii. IoT Trust Security and Privacy Model
- Reference architecture
  a. Base layer: it comprises IoT devices that have all the components like sensors with the capacity to sense compute and interface other devices.
  b. IoT gateway or aggregation layer: to gather sensor data.
  c. Processing layer: this is based on the cloud which has numerous algorithms and data processing elements.
  d. Application layer: it acts as the interface between the third-party application and use for the security of the architecture.

In this chapter, we will discuss the reference architectures of IoT in detail and also understand that how data can be collected from the environment and stored in the cloud and after that how the processing of that collected data is performed.

We know that IoT is a system where objects are being connected through the Internet and also able to collect and transfer data over a wireless network. Here is the basic block diagram of the IoT as shown in Figure 3.1, which includes the physical world and virtual world which uses a common communication network.

3.2 LITERATURE REVIEW

Table 3.1 summarizes the literature work where IoT is applied or can be applied. The discussion section of the table highlights the common and difference part of this chapter with the reference work.

TABLE 3.1
Comparison of State-of-the-Art Works for IoT Reference Architecture

Authors	Summary	Description
Verma, V., Chowdary, V., Gupta, M. K., & Mondal, A. K. (2018)	This article is all about various uses of automated frameworks, different parts of current innovations including clinical imaging, telemedicine, and supply chains have been covered concerning the COVID-19 pandemic.	This chapter has provided the idea for handling the collected data and given the brief knowledge about the reference architecture.
Chowdary, V., Kaundal, V., Sharma, P., & Mondal, A. K. (2018)	This article is based on joining of big data investigation regarding clinical practice and clinical utility, models and bunching strategies for IoT information handling, viable structures for expulsion of misclassified occurrences, common sense of big data examination, methodological and specialized issues, capability of Hadoop in overseeing medical care information is the need of great importance.	This part will help about the dealings of the big data and also explain all the layers of the architecture of IoT.
Quasim, M. T., Khan, M. A., Algarni, F., & Alshahrani, M. M. (2021)	The idea of smart cities is getting upgraded step by step. In smart cities, enormous measure of information is persistently gotten from numerous sensors, self-governing machines, or intelligent IoT gadgets. This chapter presents crucial or keen knowledge of smart cities, vertical in smart cities, and information examination approaches.	For controlling the smart cities easily, this chapter has provided the information about the method of collecting and shearing of data through the layers of IoT architecture.
Alam, T., Khan, M. A., Gharaibeh, N. K., & Gharaibeh, M. K. (2021)	A few nations are intending to receive the idea of smart city in their towns and implementing huge scope information projects that elevates smart city highlights to accomplish the prescribed degree of feasible improvement to upgrade the personal satisfaction.	For making the concept of smart city successful, this part has given the knowledge about the infrastructure of all the layers of architecture of IoT and also provided the working principle of each and every layers.
Abuhasel, K. A., & Khan, M. A. (2020)	The industrial Internet of things (IIoT) can change the current separated mechanical framework to an associated network. The deployed sensors in IIoT screen the states of the modern gadgets and machines. Consequently, unwavering quality and security became the main worries in IIoT. This paper has discussed about a safe undertaking booking framework that is given as an option in contrast to the recently applied calculations. This calculation utilizes the SoftMax-DNN and improved RSA procedures for IIoT applications.	This chapter has discussed about the way of connecting the sensors in gadgets for making a smart device and also elaborated the communication part.

(continued)

TABLE 3.1 (*Continued*)
Comparison of State-of-the-Art Works for IoT Reference Architecture

Authors	Summary	Description
Khan, M. A., & Abuhasel, K. A. (2020)	The heterogeneous industrial Internet of things (HetIoT) is a new region of exploration that can change both our impression of central informatics and the adequacy of future machines. To depict the proficiency of refreshed HetIoT machines, this article presents the idea of advance machine-metameric dimension (AmD) to examine efficacy, productivity, and adequacy.	For making a system more accurate and efficient, we need to handle the process of collecting and shearing data more carefully so this chapter has given a review about how the IoT works precisely to the present status of the IoT reference architecture.
Kumar, S., Tiwari, P., & Zymbler, M. (2019)	This article provided a brief introduction of IoT and its application in real world and also discussed about the Big data and its analysis with respect to IoT.	IoT is a system where objects are being connected through Internet and also able to collect and transfer data over a wireless network. This chapter has provided a brief knowledge about the working principle of IoT through its architectural layers.
Khajenasiri, I., Estebsari, A., Verhelst, M., & Gielen, G. (2017)	This paper gives an overview of the flexible IoT hierarchical architecture mode and how the energy will be controlled for smart cities. It gave the idea about the future problems and solutions of IoT.	This chapter helps in handling the collected data and also provides a brief knowledge about each and every parts of the layer of IoT architecture for controlling the whole process.
Alavi, A. H., Jiao, P., Buttlar, W. G., & Lajnef, N. (2018)	This paper provided information about the surface water quality monitoring network (WQMN), which is an important challenge in the smart city and also provided the design methods and optimization objectives.	This chapter has provided the details about the IoT and more precisely the architecture of IoT. With the help of this chapter, the process of surface water quality monitoring network would be improved.
Madsen, H., Burtschy, B., Albeanu, G., & Popentiu-Vladicescu, Fl. (2013)	This article gave the idea about the cloud networks and also discussed about the migration of data between two storages especially between cloud and fog layer which is called ADAS mechanism.	This part has discussed about all the layers of architecture very preciously and also the process for controlling the migration of data between all the layers. It also describes the whole model of architecture of IoT in details.

3.3 INTERNET OF THINGS (IoT) REFERENCE ARCHITECTURE

The reference architecture as shown in Figure 3.2 covers different viewpoints of cloud architecture that permit us to screen, oversee, collaborate with, and measure the information from the IoT gadget [1]. The prerequisites are anything but difficult to get a handle on connectivity and interchanges are important. This may include information assortment and information spread to different layers of architecture.

IoT is getting quicker, depending on different application regions. It works as required to and according to the plan. It contains a characterized fabrication that is carefully followed all around. The engineering of IoT relies on its value and utilization in various areas. All things considered, there is an essential cycle stream dependent on which IoT is assembled [2–4].

The elevated level of IoT reference architecture layers is as follows:

- External correspondences
- Experience examination (counting information stockpiling)
- Aggregation/bus layer – ESB and message representative
- Device interchanges
- Devices

Cross-cutting layers include:

- Device and application the board
- Identity and access the executives
- **Things** incorporate a wide range of existing gadgets that can create information straightforwardly or by changing a simple gadget signal into valuable computerized information. Moreover, inside the environment, things are virtual duplicates of articles comprised of various types of information [5–7].

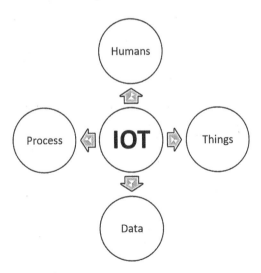

FIGURE 3.2 IoT reference architecture.

- **Data** are the computerized portrayal of the different actual structures, protests, and even personalities accordingly empowered to be interchangeable, handled by machines, and changed over into helpful data utilized by people as an intelligent decision.
- **Humans** as end-users and associated with the Internet become functioning entertainers of the IoT environment by utilizing a cycle of information delivering, changing, and sharing.
- **Processes** as a typical availability and correspondence descriptor lead us to the generally IoT environment which can ingest information self-sufficiently and go about as much as self-arranged structure.

By speaking to actual items and cycles by their virtual duplicates, the IoT biological system is restricted by a degree of discontinuity. Since this is an inconsistency, even the environment should be defeated somehow or another. It is additionally important to permit unhindered information stream and, along these lines, the assessment of the occasions in the arrangement of autonomous dynamic. For this, it is important to characterize the structure, for example, the principles of conduct and working of the components inside the IoT environment [8–10]. Since the entirety of this as of now harmonizes and is practically communicated through previously existing mechanical hypothesizes, it is least demanding to characterize this through an exceptional, for example, the reference design of the IoT biological system.

3.3.1 Prerequisites for Reference Architecture

Some key points of prerequisites for reference architecture are:

- Connection and communication
- Gadget management
- Security
- Analytical examination
- Combination

3.3.1.1 Connection and Communications

IoT connections are strategies of correspondence that ensure security to the data being exchanged between IoT-associated devices. The IoT gadgets can be associated through an IP organization or a non-IP organization [11, 12].

3.3.1.2 Gadget Management

While numerous IoT gadgets are not effectively dealt with, this isn't perfect. Nowadays, active managements are performed in PCs, cell phones, and different gadgets. What are the necessities for IoT gadget management is discussed below?

- Capacity to detach a rebel or taken gadget
- Capacity to refresh the product on a gadget
- Update security qualifications

- Remotely empowering or impairing certain equipment capacities
- Find out a lost gadget
- Wiping secure information from a taken gadget
- Remotely re-designing Wi-Fi or organization boundaries

The rundown isn't comprehensive, and alternately covers angles that may not be needed or workable for specific gadgets.

3.3.1.3 Security
The most important part of IoT is security. IoT gadgets are regularly gathering exceptionally close-to-home information. In this way, it brings the real world inside the virtual world [13, 14].

Different types of risks are:

- Risks can be present in any Internet framework, but the IoT designer may not be aware of them.
- Exact risks can be present to IoT gadgets that are out of the ordinary.

The security prerequisites in this manner should uphold

- Encryption on gadgets is sufficiently amazing.
- An advanced personality model dependent on tokens and not client IDs and passwords.
- Strategy-based and client oversaw admittance control for the framework dependent on XACML.

3.4 THE LAYERS OF THE IoT ARCHITECTURE

3.4.1 THREE-LAYER AND FIVE-LAYER ARCHITECTURES

The most basic design is a three-layer design as shown in Figure 3.3. It was introduced in the beginning phases of the examination. The three layers are the perception layer, network layer, and application layer.

- The **perception layer** is the physical layer where sensors are present for collecting the information from the environment and it also senses some physical parameters and identifies other smart devices present in the environment.
- The **network layer** is mainly used for transmitting and processing sensor data. This layer established the connection between other smart devices.
- The **application layer** provides the application-specific services to the user. It also provides various applications for developing the IoT technology.

The three-layer design characterizes the basic idea of the IoT, but it is not sufficient for IoT. That is the reason, we have a lot more layered designs proposed. One is the five-layer design, which is in addition to the preparing and business layers.

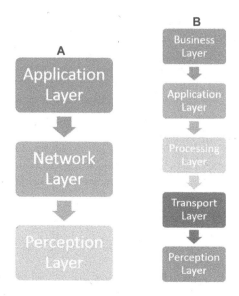

FIGURE 3.3 Architecture of IoT; (A) three layers; (B) five layers.

The five layers are perception, transport, processing, application, and business layer (see Figure 3.3). The working function of this layer is similar to the three-layered architecture.

- The transport layer sends the data of the sensor from the perception layer to the processing layer and vice versa.
- The processing layer is also called the middleware layer. It stores the data, investigates, and measures the information that comes from the transport layer.
- The business layer manages the whole IoT system.

3.4.2 Seven-Layer IoT Architecture

Seven layers of IoT design as shown in Figure 3.4 are the ones most usually utilized by clients (alluded by) when endeavoring to clarify IoT biological system appearance and its structure.

The things – to acknowledge one IoT climate, for example, the environment needs to have an assortment of gadgets, sensors, and regulators that empower their interconnection. This additionally implies that the end purpose of an IoT framework is probably associated with gadgets – other than standard sensors, the actuators likewise incorporate cell phones, miniature regulators, PCs, and so on.

Layer two – **Connectivity/Edge** speaks to the environment and where all associations are made before the trading of information inside the IoT environment. It characterizes all correspondence conventions and builds up an organization for Edge figuring. Consequently, this kind of design reveals to us that this is dispersed engineering and that information are prepared on the edge of the organization. The third layer, the **worldwide** framework, is typically a layer that depends on the cloud

IoT Solution Reference Architectures

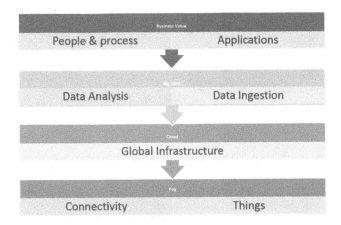

FIGURE 3.4 Seven layers of IoT architecture.

foundation. This is because most IoT arrangements depend on the mix of cloud administrations. Seen from the business viewpoint, this is an inescapable arrangement as of late because the cloud gives a total move up to the client's point of view.

The fourth layer, the **data ingestion**, is the information section layer. This is unavoidably considered Big Data, just as the purifying and information store. Additionally, information streaming cycles are available in this layer as a structural component of information ingestion.

Layer five has a place in the **data analysis** or **information examination**. This is identified with the handling of information to set up the report, information mining, the execution of AI, and so on. In the sixth layer, the **supposed application** layer, client applications are put away relies upon the reason and needs of the client. Diverse setting components are likewise thing-explicit and it is consequently important to have a characterized application layer, else they can't be normalized inside the IoT environment. Then again, this implies that this layer is the place where the incorporation of clients and items from the last layer of design happens. In certain examinations on IoT engineering, this layer is likewise called the application reconciliation layer where a similar layer can be seen as a service layer with the execution of the UI at the top.

The seventh layer is spoken to by individuals and cycles or **people** and **processes**. This incorporates all business elements as a consortium of IoT environments and, simultaneously, the entertainers associated with dynamic based on information got from the IoT biological system, with the assistance of the relative multitude of structures that were recently referenced in design [15].

3.5 IoT ARCHITECTURE BASED ON FOG COMPUTING PARADIGM

Fog layer design has a more favorable position. Quite possibly the most significant is an enhancement, it adds a unique note in a critical preferred position in the continuous exhibition of the framework when all is said in done. Figure 3.5 shows the architecture of fog layer design.

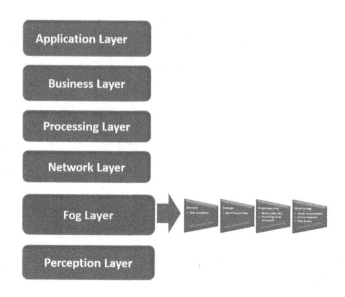

FIGURE 3.5 Fog layer architecture.

One more of the intriguing prospects that give the presentation of the fog layer in the IoT system is that the fog layer can have direct communication with another fog layer and in this way, it creates a mesh that avoids the use of cloud resources.

3.6 EDGE COMPUTING IoT ARCHITECTURE

Edge computing is firmly connected with the previously mentioned fog computing. The principal objective of edge is that the information handling and practical preparing abilities are diverted to the edge components of the organization climate inside the IoT environment [12, 16–22].

Inside the edge environment, all information preparing happens on the actual discernment layer itself, or straightforwardly on a keen gadget or an IoT gadget gatherer. Every one of these edges inside the edge can be performed autonomously on their layer or in blend with other fog or edge layers from the environmental factors of the IoT biological system.

As a significant advantage, edge permits a more prominent degree of confined idleness decrease than any past one. This is additionally encouraged by the likely decentralized association of the IoT biological system components in a worldwide environment. Simultaneously, the protection of information is ensured by the framework, and clients and internationally expand the security of the IoT biological system [23].

3.7 HYBRID IoT ARCHITECTURE

Contingent upon the undertaking necessity, the design of the IoT framework can likewise be established as a combination of Cloud-Fog-Edge. This is especially profitable with regards to testing business objectives where it is hard to meet

IoT Solution Reference Architectures

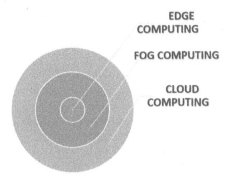

FIGURE 3.6 Hybrid IoT architecture.

clients' necessities for a portion of the standard structures. This blend, for example hybridization, is typically called nested. Architecture of hybrid IoT is as shown in Figure 3.6.

3.7.1 Edge Computing Layer

It performs perception and recording client collaborations and advances the feed to the Fog hub. From the continuous control signals from the Fog Nodes, the knowledge of the activity is performed straightforwardly at the hub level.

3.7.2 Fog Computing Layer

All current information is put away in impermanent memory. The control and investigation needed to run progressively depend on the application center guidelines from the cloud.

3.7.3 Cloud Computing Layer

It performs accumulation of information from all Fog hubs and performs scientific cycles on huge datasets. Moreover, it has the assignment of sending rules for applications execution of Fog Nodes.

3.8 SOCIAL IoT

Social IoT or SIoT is a new paradigm. In this part, we maintain the relationships between objects like humans have. Here we will discuss the characteristics of SIoT.

- SIoT acts as a navigator; it navigates all the devices considering one starting device and establishes the social network. By this process, we can easily identify new devices.
- SIoT also maintains security between two partners and makes their bonding strong enough.

3.8.1 Basic Components

Basic component is the part of the social IoT, where we treat the gadgets and organization as bots. They can set up connections among them and adjust them after some time. This will allow the gadgets to coordinate with each other. To make a particular model work, we need to have various interoperation sections.

- **ID:** ID is an identity that provides the recognition conformation.
- **Meta information:** Along with an ID, this part also gives the information about information.
- **Controls security:** This is like "friend list" settings on Facebook. An owner of a gadget has the power to decide that what kind of devices can be connected to it. These are mainly controlled by the owner.
- **The revelation of service:** This system is similar to the service cloud. Where we have many directions which store details of devices from different services. We need to store these directions and make them updated so that devices can learn from each other.
- **Connection to the executives:** This part establishes the relation between different devices. It also stores information about the services of other devices.

3.9 CONCLUSION

The IoT is influencing our lifestyle from the way we react to the way we behave. We can control our AC through our smartphone, our smartwatch always tracks our daily activity, and smart cars provide the shortest route. These all are examples of IoT. IoT is a giant network that connects all the devices, these devices gather data and share data about how they are used and the environment in which they are operated.

Today, we can predict that in the future IoT our world is going to be more like the movie of blatant and ironman kind where everything we used is connected to the Internet, where all of our data is being collected and being utilized to make our life and society a much better place. We can assume that almost everything is going to be automatic and connected to the Internet like our table, chair, doors, hangers, and many more.

This part has given a review about how the IoT works precisely in the present status of the IoT reference architecture; however, this chapter has covered the viewpoints of IoT reference architecture that are imperative to help in the present IoT arrangements.

REFERENCES

1. Mukherjee, M., Adhikary, I., Mondal, S., Mondal, A. K., Pundir, M., & Chowdary, V. (2017). A Vision of IoT: Applications, Challenges, and Opportunities with Dehradun Perspective. In Proceeding of International Conference on Intelligent Communication, Control and Devices (pp. 553–559). Springer, Singapore.
2. Quasim, M. T., Khan, M. A., Algarni, F., & Alshahrani, M. M. (2021). Fundamentals of Smart Cities. In Khan, M. A., Algarni, F., & Quasim, M. T. (eds) Smart Cities: A Data Analytics Perspective. Lecture Notes in Intelligent Transportation and Infrastructure. Springer, Cham. https://doi.org/10.1007/978-3-030-60922-1_1

3. Alam, T., Khan, M. A., Gharaibeh, N. K., & Gharaibeh, M. K. (2021). Big Data for Smart Cities: A Case Study of NEOM City, Saudi Arabia. In Khan, M. A., Algarni, F., & Quasim, M. T. (eds) Smart Cities: A Data Analytics Perspective. Lecture Notes in Intelligent Transportation and Infrastructure. Springer, Cham. https://doi.org/10.1007/978-3-030-60922-1_11
4. Khan, M. A., Quasim, M. T, et.al. (2020). A Secure Framework for Authentication and Encryption Using Improved ECC for IoT-Based Medical Sensor Data, in *IEEE Access*, vol. 8, pp. 52018–52027, doi: 10.1109/ACCESS.2020.2980739
5. Abuhasel, K. A., & Khan, M. A. (2020). A Secure Industrial Internet of Things (IIoT) Framework for Resource Management in Smart Manufacturing, in *IEEE Access*, vol. 8, pp. 117354–117364, doi: 10.1109/ACCESS.2020.3004711
6. Khan, M. A., & Abuhasel, K. A. (2020). Advanced Metameric Dimension Framework for Heterogeneous Industrial Internet of Things, in *Computational Intelligence*, vol. 2020, pp. 1–21. https://doi.org/10.1111/coin.12378
7. Kumar, S., Tiwari, P., & Zymbler, M. (2019). Internet of Things is a Revolutionary Approach for Future Technology Enhancement: A Review, in *Journal of Big Data*, vol. 6, p. 111. https://doi.org/10.1186/s40537-019-0268-2
8. Zhou, J., Cap, Z., Dong, X., & Vasilakos, A. V. (2017). Security and Privacy for Cloud-Based IoT: Challenges, in *IEEE Communications Magazine*, vol. 55, no. 1, pp. 26–33. https://doi.org/10.1109/MCOM.2017.1600363CM
9. Khajenasiri, I., Estebsari, A., Verhelst, M., & Gielen, G. (2017). A Review on Internet of Things for Intelligent Energy Control in Buildings for Smart City Applications, in *Energy Procedia*, vol. 111, pp. 770–779.
10. Alavi, A. H., Jiao, P., Buttlar, W. G., & Lajnef, N. (2018). Internet of Things-Enabled Smart Cities: State-of-the-Art and Future Trends, in *Measurement*, vol. 129, pp. 589–606.
11. Madsen, H., Burtschy, B., Albeanu, G., & Popentiu-Vladicescu, F. (2013). Reliability in the Utility Computing Era: Towards Reliable Fog Computing. In Proc. 20th International Conference on Systems, Signals, and Image Processing (IWSSIP). pp. 43–46.
12. Mukherjee, A., Goswami, P., Khan, M. A., Manman, L., Yang, L., & Pillai, P. (2021, April 1). Energy-Efficient Resource Allocation Strategy in Massive IoT for Industrial 6G Applications, in *IEEE Internet of Things Journal*, vol. 8, no. 7, pp. 5194–5201, doi: 10.1109/JIOT.2020.3035608
13. Verma, V., Chowdary, V., Gupta, M. K., & Mondal, A. K. (2018). IoT and Robotics in Healthcare. In Medical Big Data and Internet of Medical Things (pp. 245–269). CRC Press, Boca Raton, USA.
14. Chowdary, V., Kaundal, V., Mondal, A. K., Devella, V., & Sharma, A. (2019). Internet of Things-Enabled Virtual Environment for U-Health Monitoring. In Sensors for Health Monitoring (pp. 117–133). Academic Press, Cambridge, Massachusetts.
15. Sharma, A., Jain, A., Gupta, P., & Chowdary, V. (2020). Machine Learning Applications for Precision Agriculture: A Comprehensive Review. in *IEEE Access*, vol. 9, pp. 4843–4873.
16. Quasim, M. T. *et al.* (2021). Emotion-Based Music Recommendation and Classification Using Machine Learning with IoT Framework, in *Soft Computing*, vol. 25, pp. 12249–12260. https://doi.org/10.1007/s00500-021-05898-9
17. Munusamy, A. et al. (2021). Service Deployment Strategy for Predictive Analysis of FinTech IoT Applications in Edge Networks, in *IEEE Internet of Things Journal*, doi: 10.1109/JIOT.2021.3078148
18. Xu, L., Zhou, X., Khan, M. A., Li, X., Menon, V. G., & Yu, X. (2021). Communication Quality Prediction for Internet of Vehicle (IoV) Networks: An Elman Approach, in *IEEE Transactions on Intelligent Transportation Systems*, doi: 10.1109/TITS.2021.3088862.

19. Verma, S., Kaur, S., Khan, M. A., & Sehdev, P. S. (2021, April 1). Toward Green Communication in 6G-Enabled Massive Internet of Things, in *IEEE Internet of Things Journal*, vol. 8, no. 7, pp. 5408–5415, doi: 10.1109/JIOT.2020.3038804
20. Munusamy, A. *et al.* (2021). Edge-Centric Secure Service Provisioning in IoT-Enabled Maritime Transportation Systems, in *IEEE Transactions on Intelligent Transportation Systems*, doi: 10.1109/TITS.2021.3102957
21. Nandy, S., Adhikari, M., Khan, M. A., Menon, V. G., & Verma, S. (2021). An Intrusion Detection Mechanism for Secured IoMT Framework Based on Swarm-Neural Network, in *IEEE Journal of Biomedical and Health Informatics*, doi: 10.1109/JBHI.2021.3101686
22. Khan, W. U., Li, X., Ihsan, A., Khan, M. A., Menon, V. G., & Ahmed, M. (2021). NOMA-Enabled Optimization Framework for Next-Generation Small-Cell IoV Networks Under Imperfect SIC Decoding, in *IEEE Transactions on Intelligent Transportation Systems*, doi: 10.1109/TITS.2021.3091402
23. Chowdary, V., Kaundal, V., Sharma, P., & Mondal, A. K. (2018). Implantable Electronics: Integration of Bio-Interfaces, Devices and Sensors. In Medical Big Data and Internet of Medical Things (pp. 55–79). CRC Press, Boca Raton, USA.

Part II

Simulation, Modeling, and Programming Framework

4 Hardware Architecture of IoT and Wearable Devices

Manoj Sharma
Bharati Vidyapeeth's College of Engineering
New Delhi, India

CONTENTS

4.1 Introduction 56
    4.1.1 Internet of Things: The Backbone 56
        4.1.1.1 Wi-Fi..57
        4.1.1.2 Bluetooth..58
        4.1.1.3 Zigbee 58
        4.1.1.4 MQTT 58
        4.1.1.5 OPC-UA..58
        4.1.1.6 Cellular..58
        4.1.1.7 Z wave 59
        4.1.1.8 NFC (Near Field Communication) 59
        4.1.1.9 LoRaWAN...59
        4.1.1.10 SigFox 59
        4.1.1.11 DDS...60
        4.1.1.12 AMQP 60
        4.1.1.13 CoAP...60
        4.1.1.14 NB-IoT 60
        4.1.1.15 Thread 60
4.2 Wearable Technology 60
    4.2.1 Inter/Intra Communication System 64
4.3 Power Management...65
4.4 Circuits Design for Wearable Systems 67
    4.4.1 Technology...67
    4.4.2 Circuits for Wearable Technology 67
        4.4.2.1 Low-Power Techniques 68
4.5 New Circuit Level Advances and Hardware Architectures..........................70
    4.5.1 Essential Hardware Architecture of IoT-Enabled Wearable System.....71
4.6 Conclusion 72
References..72

4.1 INTRODUCTION

Wearable devices are one of the popular applications of Internet of things (IoT) which has become an integral part of the human day-to-day life. It has become ubiquitous finding usage in every sphere of life. Devices like smart watch, bit band, smart cloths, smartphones, and medical wearables have revolutionized the market. The acceptance of wearable devices among the masses has brought IoT inevitably deeper into the lifestyle. The convergence of physical and digital world through wearable devices has surely improved the human life standard in health, fitness, safety, identity checks, interactions, and many more. In coming years, market would have numerous products related to health monitoring, controlling objects, operations, devices like doors, and authentications-based applications. Future wearable devices must meet ever-growing expectations of the consumers. This involves new technologies in sensors, data exchange systems, and circuits used for wearable devices. This chapter deals with availability of technologies and for this focusing on communication protocols, circuit, and power requirements.

4.1.1 INTERNET OF THINGS: THE BACKBONE

IoT involves interconnection of large number of gadgets speaking with each other. These gadgets establish intercommunications in virtue of communication network. The everyday objects send and receive data in IoT network. Therefore, communication protocol is very critical and crucial for efficient implementation of IoT network. These protocols are the set of rules required for intertransferring the data among different devices. In a CAGR forecast, the IoT communication protocol market would reach to USD 15.80 billion by 2022 as compared to USD 11.44 billion in 2015 [1, 2].

Globally in last 20 years, IoT has created an interconnected network of over 20 billion gadgets. With the help of these strong interconnections, the gadgets capture data, do processing over it, and communicate with other nodes spread across different parts of world. The applications of these data-sharing devices cover sectors ranging from purely production-based industries like manufacturing, oil and gas, and utilities to service-based companies like transportation, healthcare, and hospitality.

However, it is important to note that these devices can only share data when the connection between them is secure; and that's what IoT standards and protocols are made for—to facilitate safe and high-speed data transfer among end IoT devices [3].

There are various IoT protocols that a business can leverage to develop end-to-end IoT solutions. Nevertheless, each of these standards have different well-defined attributes and competencies, due to which the selection of best suitable protocol becomes very tedious.

According to Gartner in 2020, IoT gadgets have grown to 26 billion units approximately. Statista, Lionel [4] forecast an interconnection of 50 billion IoT devices by 2030 globally. Because of this huge amount of interconnection, it is very critical to understand IoT communication protocols and standards which would benefit in

Hardware Architecture of IoT and Wearable Devices

efficient implementation and potentially reduce the security breaches at bay. The major factors assisting the growth of IoT are:

1. Availability of Internet to all populations
2. Increase in smartphone usage
3. Availability of cloud computing
4. Availability of different sensors including wireless sensors
5. Increasing technical awareness and adopting preferences among people

The smart IoT-enabled devices may have security threats which can be minimized with the use of right protocols. The IoT communication protocols are backbone of the data exchange among devices ensuring optimum data security. Some of the major advantages of using IoT communication protocols in the IoT ecosystems pertain to higher credibility, higher security, higher quality, higher opportunities for scalability, and interoperability. This also reduces the IoT system complexity at design, implementation, and maintenance levels.

These protocols may be IP network based or non-IP-based network strategies. IP-based strategies have major advantage of high range but suffer with high power and memory requirements as compared to non-IP-based systems. Further as in Figure 4.1, one can classify them as IoT network protocols which connect medium- to high-power devices. They are used to connect devices over the network/Internet. HTTP, LoRaWAN, Bluetooth, and Zigbee are some examples of IoT network protocols [1, 5].

The IoT data protocols are used to interconnect power aware IoT gadgets by providing point-to-point communication. Data exchange among the devices is possible even without Internet which may utilize wired or cellular system for the same. CoAP, MQTT, XMPP, AMQP are some of the examples of IoT data protocols. Each of the IoT communication protocols have their own advantages and disadvantages and feature sets. Wi-Fi and Bluetooth are most common protocols used in IoT applications. Details about some promising communication protocols and standards that one can use to build IoT solutions are given in the next paragraphs.

4.1.1.1 Wi-Fi

Wireless Fidelity (Wi-Fi) is a wireless LAN and utilized IEEE 802.11n standard with 2.4 GHz to 5 GHz bands. Wi-Fi is used in homes and offices having hundreds

FIGURE 4.1 IoT network classification.

of Mbps data transfer rate in a range of 50–100 m from access point. It can provide a maximum 600 Mbps but 150–200 Mbps is commonly available based upon channel frequency used and number of antennas. But the Wi-Fi-based system is very power hungry so they are not suitable for battery-operated devices. Wi-Fi systems require low infrastructure costing and are best for indoor applications [6].

4.1.1.2 Bluetooth

It is a short-range wireless personal area network communication protocol utilizing IEEE 802.15.1 standard with 2.4 GHz to 2.485 GHz. Bluetooth is used for small amount of data transfer with a data rate of 1 Mbps in a range of 50–150 m from access point. With reduced power requirements, Bluetooth is very useful for IoT systems.

The frequency band of Bluetooth is similar to that of Wi-Fi suggesting the two technologies are similar but they have their own specific usages. Bluetooth have three different variations. Traditional Bluetooth is very power hungry. The Bluetooth Low Power (BLE) or Bluetooth 4 which was introduced by Nokia is adopted by all operating systems because of its power efficiency. iBeacon variant of Bluetooth is used by Apple and is very similar to NFC.

4.1.1.3 Zigbee

It is a robust, highly secure, low power protocol based on IEEE 802.15.4 standard with 2.4 GHz. Wi-Fi is used in industrial application more with a maximum of 1024 nodes in a range of 200 m with a data rate of 150 kbps. It also has 128-bit AES encryption capability. Some of the benefits of Zigbee are low power consumption, strong security, high scalability, and durability.

4.1.1.4 MQTT

Message Query Telemetry Transport (MQTT) is a light weight message protocol. It sends data from sensors to applications and middleware. Working over TCP/IP, it can integrate with any network having ordered lossless and bidirectional connects. It utilizes ISO/IEC 20922 standard with a data rate of 256 Mb.

4.1.1.5 OPC-UA

Like MQTT, it is also a platform-independent protocol and popular in industrial applications.

4.1.1.6 Cellular

It is based on GSM/GPRS/CDMA/EDGE(2G), UMTS/HSPA(3G), LTE(4G) standard with 900/1800/1900/2100 MHz frequency in the range of 35–200 km having data rate of 35–170 kbps. This requires huge power and can support numerous amounts of data load even across continents. Many times, it is also referred as Satellite or Machine-to-Machine (M2M) communication system in IoT as it permits sending and receiving data using cellular network. This form of communication has advantages of satiability and universal compatibility. But the huge monthly running cost is another disadvantage of cellular system.

Hardware Architecture of IoT and Wearable Devices

4.1.1.7 Z wave

It is based on power aware RF-communication technology and best used for home automation-related applications where devices communicate in the range of 25–40 m with a data rate of 0.3–50 kbps connecting up to 232 devices. It is based on Z wave alliance standard. With sub-1 GHz (900 MHz) band it provides lowest latency among all IoT protocols. It may also be made compatible to cloud storage.

4.1.1.8 NFC (Near Field Communication)

NFC is based upon ISO/IEC 18000-3 standard in the range of 10 cm working on 13.56 MHz frequency with data rate of 100–420 kbps. It is based upon the interactions in electromagnetic radiations among small-loop antennas (forming an air core-transformer) in very close proximity and widely used in smartphone and digital payment applications. NFC initiator generates RF field which provides power to the passive tag, sticker, key fob, or any battery less card named as passive target. BLE and NFC can be compared as below:

Parameters	BLE	NFC
Speed	faster	slower
Transfer	Higher rate 2.1 Mbps	Slower rate 424 kbps
Power	Consumes more power	Consumes less power
Pairing	Requires	Does not require
Time to setup	Takes more	Less time
Connection	–	Automatically
Range	Comparatively larger	Comparatively shorter
Compatibility	–	With existing passive RFID

4.1.1.9 LoRaWAN

Incepted in 2015, this is a two-way, long-range radio WAN providing cost-effective mobile security for IoT targeting smart cities, smart towns, smart villages, and many industrial applications which can support millions of devices with data rate of 0.3–50 kbps having a range of 2.5–15 km. Also, it is suitable for battery-powered devices [7].

4.1.1.10 SigFox

Working at frequency 900 MHz with data rate of 10–1000 bps SigFox aims low-cost M2M applications. It has capabilities of cellular and Wi-Fi protocols. This is quite useful for battery-operated devices. It finds applications ranging from home appliances, consumer goods, retail industry like PoS, etc.

Zigbee and Z wave use a low-power RF radio frequency bands which are country specific. India has 865.2 MHz band; Europe has 868.42 MHz SRD band. RFID is another popular technology based on electromagnetic field. Active Reader Passive Tag (ARPT) system and Active Reader Active Tag (ARAT) system find variety of application in IoT like animal identification, factory data collection, road tolls, building access, pharmaceuticals, and other inventory monitoring systems.

IoT protocols have three basic layers which are application, network, and perception. Sometimes they are extended to five layers including the business and processing

TABLE 4.1
Classification of Protocol

Short Form	Usage
TCP	For data exchange
FTP	For transferring digital files/data from client-server interconnected system
UDP	For sending short-datagram-messages
HTTP	Application protocol, involving hyperlink- connecting nodes
ICMP	For diagnostics
POP	Mails from remote TCP IP-based server system to local email clients
IMAP	Mail server: storage emails

layers. MQTT and CoAP are two lightweight protocols which are leading messaging protocols. Advanced Message Queuing Protocol (AMQP) is an open standard IoT protocol. The data exchange protocols can also be classified as given in Table 4.1.

The selection of protocol acts as one of the important factors in deciding the volume, speed, and range of data transmission [1, 2, 8].

4.1.1.11 DDS
DDS can transfer data to cloud platform and to low-footprint devices making use of DCPS and DLRL layers. It is a high-performance real-time M2M and expandable IoT standard.

4.1.1.12 AMQP
It utilizes message-oriented architecture for data transfer and is designed mainly for middleware environments.

4.1.1.13 CoAP
Constrained Application Protocol (CoAP) makes use of HTTP and user data protocol (UDP) for lightweight data with restful architecture and binary data formats. It is best suited for applications like automation, mobile phones, and microcontrollers.

4.1.1.14 NB-IoT
Narrow Band IoT (NB-IoT) is designed for low-bandwidth, low-latency applications supporting massive connection density. It also supports data security features and consumes least power.

4.1.1.15 Thread
It is a low power mesh networking protocol which is based on IEEE 802.15.4 radio standards. It has inherent performance capability for scaling the mesh. A summary comparison table regarding some popular standards is shown in Table 4.2.

4.2 WEARABLE TECHNOLOGY

Wearable is a subset of IoT and many a times wearable technology and wearable devices are interchangeably used. They primarily refer to the application of electronic

TABLE 4.2
Comparison Table

Protocol	Managing Body	Range	Data Rate	Frequency	Topology	Open/ Proprietary	Remark
Zigbee	Bluetooth Special Interest Group (SIG)	850 ft	250 kbps	2.4 GHz	Mesh	Open	Low power
Thread	Thread Group (Google, Samsung, etc.)	200 ft					
BLE[a]	Bluetooth Special Interest Group (SIG)	300 ft	125 Kb/s to 2 Mb/s				
LoRa	LoRa Alliance	9 miles	50 kbps	150 MHz– 1 GHz	Star		
NB-IoT	3GPP, Ericsson, Huawei	18 miles	100 kbps	<1 GHz			
Wi-Fi	IEEE	200 ft	7 Gbps	2.4 GHZ/ 5 GHz			High power

[a] P2P topology also.

technologies embedded with wearable accessories like clothing, ornaments, or specially designed wearable pieces. These may additionally also consist of computational capabilities along with electronic circuits. Wearable devices are more complex as compared to simple prevalent electronic circuits as it has to deal with biometric parameters of human and animals. The ultimate aim of the wearable technology is to provide the benefit of electronic circuits in a convenient, seamless, portable, and mostly hand-free access targeting health, communication, and enjoyment-related applications [9, 10]. Therefore, one can infer that it has intelligence which accepts input from multiple sensors, processes these inputs, and provides a meaningful output. It inherently involves miniaturization of integrated circuits and their interconnections. Some of the areas of applications of wearable technology are illustrated in Figure 4.2.

The key parameters that are associated with wearable technology are:

1. Functionality metrics: It aims the vital signs to be monitored, interconnection with external devices, safety, and quality of care.
2. Performance metrics: It involves the parameters like product size, data rates, data latency, fault tolerance, data security, cost, manufacturability, etc.
3. Wearability: This involves comfortability level associated with the wearable device. It also involves parameters like ease of access, ease in motion after wearing it, friendliness to skin, moisture absorption, etc.

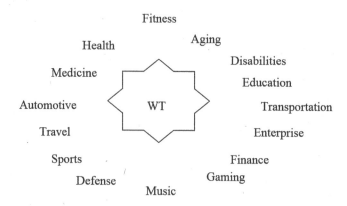

FIGURE 4.2 Areas of applications of wearable technology [11].

4. Maintainability: This covers the aspects of scaling, software upgradation, rechargeability, cleaning, etc.
5. Field usability: It deals with efficient electrostatic charge protection, safety from EMI radiations, usage of standard protocols for hardware, software design, and intercommunication. It also involves protection from biomechanical aspect along with plug and play capability.
6. Durability: It deals with parameters like heat abrasion, corrosion resistance, and mechanical strength.

The data transfer network in a wearable device constitutes body area network (BAN). A typical wireless BAN architecture is demonstrated in Figure 4.3.

The first layer involves an intra-WBAN communication which involves collation of sensor's data at one centralized hub at the body/wearable device with the subject [12]. The data collected is sent to intermediate inter-WBAN level from where data can be fetched globally utilizing Internet services. The cellular system is most

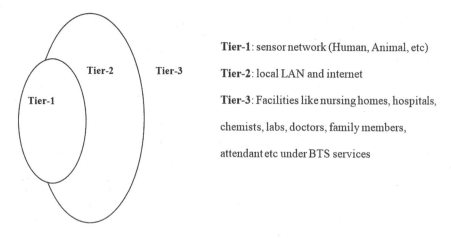

FIGURE 4.3 WT communication layers.

TABLE 4.3
WBAN Applications

Human and Animal Wearable Networks (BAN) and Devices

Wireless	Implantable	Remote Controlled	Non-medical
Assessing soldier fatigue	Cardiovascular illness	Patient monitoring	Entertainment
Assessing driver fatigue	Cancer detection	Tele-medicine solutions	Natural calamities
Assigning employees attentiveness	Blood pH levels	Ambient-assisted living	Societal emergencies
Sleep staging			
Illness like asthma, heart related, temperature, etc.			

widely used network in inter-WBAN network. In the third layer, the data is sent to the desired authentic nodes in a secured manner. Some of the applications of WBAN are listed in Table 4.3.

The signal propagation in the body is affected due to electrical properties of the body, as summarized in Table 4.4. The body tissues can absorb some part of the signals as they are semi-conductive. The dielectric constants εr, conductivities ρ, characteristic impedances $Z(\Omega)$ are different for muscle and fat. As per the frequency there occurs a certain loss in human body.

Following are the key design parameters that should be considered at the planning for a wearable device:

- Technical architecture: open, modular, and flexible
- Operating environment: operate in targeted environment, like high temperature, under rivers/ocean, ice-bergs, etc.
- Minimizing radiations: minimize EMF radiations

Further every wearable device constitutes the following components:

- Sensor network: to detect desired parameter system can have one or more sensors

TABLE 4.4
Characteristics of In-/On-Body Applications [12]

Human Body Impedance Z_o (Ohm)

Frequency (MHz)	Muscle	Fat
100	31.6	92.4
400	43.7	108
900	48.2	111

- Data processing: system needs a microcontroller and application software to extract data and make intended/intelligent use of data coming from designed sensor networks
- Communication network: system requires communication channel to interchange data among devices which may be inside the system and/or to the outside world utilizing wired or wireless communication network technologies
- User interface: input devices like keyboard, sensor-network, touch devices, voice/speech, and the output devices like LED display, printer, messages, out-speech, etc. are very critical to provide a good user interaction mechanism

For designing an efficient wearable system, following components become critical:

- Inter/intra communication system
- Power management system
- Circuits design

4.2.1 Inter/Intra Communication System

As depicted in Figure 4.4, every wearable device consists of three layers of communication each supporting the security aspects as well. It is always preferable that all communication protocols used for inter or intra communication should use same message/data structure. Wired connections of these sub-modules have certain limitations. A Body Area Network (BAN) or a Body Sensor Network (BSN) are generally implemented. Bluetooth and Zigbee are very popular in BAN. The data format should have header, message body, and tail.

- Header: source, destination, type of content, etc.
- Body: exact sensed content, analyzed results
- Tail: Error correction capabilities

The parameter details for some of the communication protocols are summarized along with approximate costing. The Zigbee standards are used for low-cost solution where data rate is low with multiyear battery life. Another critical design parameter favoring the usage of it is very low complexity involved in system design and implementation. The system mostly uses 16 channels in ISM band with OQPSK modulation or BPSK for 10-channel system. Further the system prefers CSMA-CA and DSSS coding. The communication network architecture can be implemented in either of tree, star, mesh, or cluster topologies. Data rate and power requirements are critical in selecting the protocols as summarized in Table 4.5 [13].

FIGURE 4.4 Wearable technology—communication network.

TABLE 4.5
Protocols, Data Rate, and Power Requirements

	Range	Data Rate	Power	Frequency
Zigbee	10–75 m	20 kbps	30 nW	868 MHz
		40 kbps		915 MHz
		250 kbps		2.4 GHz
Bluetooth	10–100 m	1–3 Mbps	2.5–100 mW	2.4 GHz
IrDA	1 m	16 Mbps	–	Infrared
MICS	2 m	500 kbps	25 uW	402–405 MHz
802.11 g	200 m	54 Mbps	1 W	2.4 GHz

Frequency hopping technique (FHSS) with 79 ISM-based channels supporting 3 Mb/s connected in the piconet, a star topology network, is popular for Bluetooth. Progressively adopting Wi-Fi PHY/MAC-layers BLE for ultra-low power applications forms the BAN system. IrDA, medical-implant-communication-service (MICS), ultra-wideband (UWB) further revolutionize the short-range BAN system in wearable devices. Seeing practical application, many a times researchers highlighted the issues in existing communication technologies for efficient BAN systems may be because of signal interference issues, power issues, security concerns, or form factor of hardware circuits.

- Adaptive data bit-rates ranging from 1 kb/s to Mb/s
- 2 m to 5 m short distance protocol
- Network of maximum 100 nodes
- Low latency level
- Power consumption in the range of 0.1–1 mW

To avoid interferences from environment, medical BAN services are allotted a 2360–2400 MHz band.

4.3 POWER MANAGEMENT

Whether for BAN or WBAN, power is one of the major issues specifically to implantable wearable devices. Figure 4.5 depicts a comparative view of the power requirements for different communication protocols and data rates. With the advances in technologies, wireless sensors generate wide range of data bits per second. A pacemaker typically generates in few Kbps and an endoscope may generate few Mbps. The traditional battery (lithium ion) system has certain limitations in the driving capabilities and flexibility. New advances have created bendable, flexible battery which surely benefits the wearable domain. A new proposed laminated outer body with internal battery structure assists in leak proof and prevents from overheat. Panasonic in its first bendable created 0.02 inch thick, bendable with a radius of

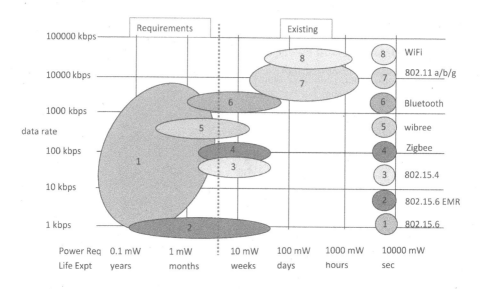

FIGURE 4.5 Data rates and power requirements [15].

25 mm, 25 degrees. In development stage the bendable battery has a 17.5 mAh to 60 mAh as compared to 1960 mAh in traditional batteries [14].

Energy harvesting is another way in efforts to address the power issue in the wearable devices. It is a process of utilizing the other nontraditional power sources like solar, wind, thermal, salinity gradient, or kinetic energy which can be captured and stored. Efforts are applied to capture the ambient energy to drive the electronics in wearable systems. A block level representation for such harvesting system is shown in Figure 4.6.

More advances are in place toward wireless charging integration. Dima et al. proposed reconfigurable architecture targeting multiple voltage levels (0.6 v, 0.8 v, and 1 v) switched-capacitor dc-dc buck converter. This proves to be one of very useful power management block for wearable electronics circuit components and systems. At 65 nm CMOS tech node the architecture based upon PFM offers a current range of 10–800 µA. Farzanah also proposed curved electromagnetic energy harvesting

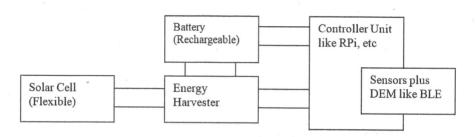

FIGURE 4.6 Energy harvesting system representation [16].

system which capture energy from oscillatory motion while walking/running in both kind of oscillations. The proposed technology is able to produce a no-load voltage from 2.55 v to 4.07 v [3, 17–21].

4.4 CIRCUITS DESIGN FOR WEARABLE SYSTEMS

4.4.1 Technology

With the advances in technologies, especially MEMS, the availability of sensors has flooded in the market. This is illustrated in Figure 4.7. In an estimation, around 300 million sensor devices have been shipped in 2020 as compared to 3.5 billion in 2014. These sensors measuring different parameters like movement, position, temperature, activity, and vital signs are backbone of the wearable devices.

4.4.2 Circuits For Wearable Technology

Wearable devices consist of other electronic circuits as well. The power efficiency for these is also very critical. Following are the key methods that can be used to develop power aware circuits:

- Playing with factors as per equations to trade off the design parameters
- Innovative design architectures
- Innovative coding techniques
- Component re-use

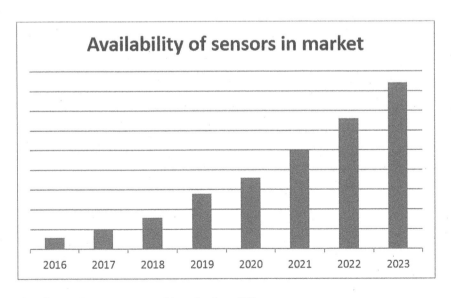

FIGURE 4.7 Sensors for wearable technology [22].

Some low-power circuit design techniques are listed as below. These techniques are well established in chip design domain for addressing the circuit power dissipation.

4.4.2.1 Low-Power Techniques

Some of the popular low-power circuit design techniques are list below:

- Active substrate/body biasing
- Adaptive voltage scaling
- Clock tree gating
- Dynamic voltage scaling
- Dynamic voltage and frequency scaling
- Restructured logic implementation
- Multiple voltage levels
- Multi-threshold (Vth) gates and devices
- Power gating
- Power gating with retention
- Scaled transistors
- State retention power gating
- Save and restore power gating
- Transistor stack
- Multiple oxide
- Multiple channel length
- Multiple body bias
- Architecture-driven voltage scaling
- Operation reduction and/or substitution
- Pre-computation

But the stated techniques are approaching their physical limitations [23]. Hence, new power aware techniques from other domains are also used. Adiabatic logic is one of such technique which shows promising opportunities for circuit development for wearable devices. In simple language, two rules can be followed to develop adiabatic logic circuit; these are:

- If potential difference between source and drain, do not turn on transistor.
- Should not turn off a MOS switches in presence of current.

Some adiabatic logic families are ECRL, PFAL, IECRL, IPFAL, EEL, PAL, and SCRL. The governing energy relation for adiabatic circuit is

$$E = \int_0^T R \frac{C^2 V_{DD}^2}{T^2} dt = \frac{RC}{T} CV_{DD}^2 \qquad (4.1)$$

Therefore, for T > 2 RC the energy is less than the conventional circuits. The author has proposed a new CPLAG circuit, RCPLAG circuit, and modified an existing PFAL for designing low power circuits targeting wearable devices. Some basic circuit modules are illustrated in Figures 4.8–4.10.

Hardware Architecture of IoT and Wearable Devices

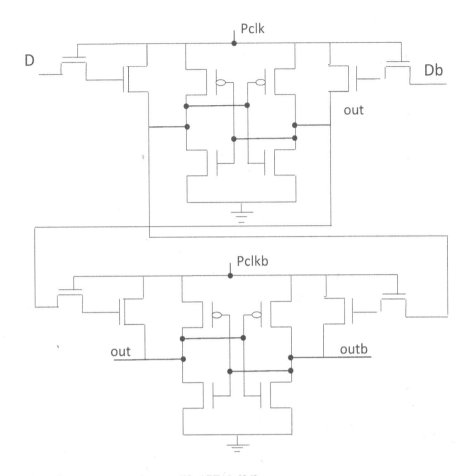

FIGURE 4.8 DFF based on modified PFAL [24].

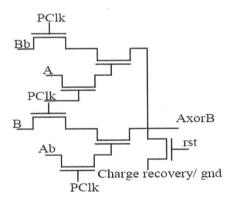

FIGURE 4.9 Circuit topology for CPLAG XOR gate [25].

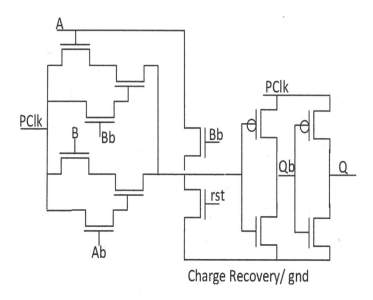

FIGURE 4.10 MCPLAG DPET DFF with adiabatic "xor" [26, 27].

4.5 NEW CIRCUIT LEVEL ADVANCES AND HARDWARE ARCHITECTURES

New advances in circuit implementation sciences have enabled Hybrid Flexible System design technique. The flexible circuit may be organic-polymer based or they can be carbon nanotube film based or inorganic semiconductor/nanowires based. These flexible electronic circuits make use of either Ink-jet printing, Gravure, Flexo, Screen Printing, Nano-imprinting or Laser-based printing approaches. Figure 4.11 demonstrates a typical view of flexible electronic circuit system. The technique is quite simple, cost effective, and can be performed in room temperature. Electronic papers, printed circuits, electronic displays, and bioelectronic circuits are major advantageous applications related to flexible electronics [28].

Ramesh et al. have proposed a cost-effective fabrication technology for radio-frequency microelectromechanical systems (RF MEMS). This technology uses flexible circuit film and movable switch membrane. The technology proves to be boon for integrating printed circuit and antennas on laminate substrate [29, 30].

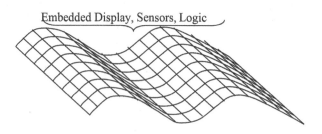

FIGURE 4.11 Flexible technology circuits [29].

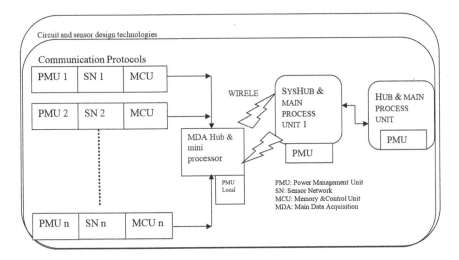

FIGURE 4.12 Hardware architecture of IoT-enabled wearable system.

4.5.1 Essential Hardware Architecture of IoT-Enabled Wearable System

Figure 4.12 depicts the proposed hardware architecture for IoT-enabled wearable systems. The architecture involves different set of sensor networks that can be employed to sense and accumulate different data sets. The sensor networks can ley on analog or digital domain and accordingly only their interface with MCU need to be designed and proper protocol may be selected as per the volume of data, range, and power availability. Each SN can have their own specific PMUs to address the power requirements of the respective sensor networks. The PMU may have static battery systems which can be replaced time to time or they can also employ rechargeable capabilities. The PMU can also use energy harvesting techniques to harvest/extract power energy from ambient. Solar and EMF radiations are most common energy harvesting mechanism that found applications in PMU with sensor networks. Further each SN unit can have its own specific memory and control unit capabilities. The intelligence embedded at SN level is useful for filtering the raw data generated by the sensors. It reduces the data bandwidth requirements and some local decision may be incorporated. The low lever controller, capabilities assist in local decisions, for effective implementation and maintenance of the sensor networks, and their interface with data acquisition unit. The MDA hub involving microcontroller or processor, provides major decision-making capabilities to maintain the SN and data transfer to wireless mechanism. This allows the information formulation and data propagation in an intelligent manner. The main processor unit augments the system with different interface capabilities and report generations. Different triggers can be initiated to take corrective actions as per information received by the main processing units. It helps to cross all the physical boundaries across globe to shape a real integrated system.

Circuit design, power management, and communication channel (inter and intra both levels) are most critical design parameters. The communication sub-system must follow intelligent data format comprising header, message body, and tail aiming least activity factor and efficient data transfer capabilities. It may involve data encoding techniques for optimizing the said two-dimensional parameters. Communication sub-system need to target mainly on transferring sensed signals to MCU and MDA hub and to integrate with wireless communication channels as per application. It may make use of wired network but would have certain constraints. Flexible smart textile-based clothing, BAN, BSN, WBAN (IEEE 802.15.6) would be attractive options making use of Zigbee (IEEE 802.15.4), Bluetooth (IEEE 802.15.1), Wi-Fi (IEEE 802.11 a/b/g/n, IrDA), medical implant communication service (MICS), UWB (IEEE 802.15.4.a). Secondly, WLAN, GSM, GPRS, UMTS, WiMAX, etc. are handy for wireless integration.

The proposed system architecture contain two main subsets. In the first part, it may use any software application language or mobile development language for efficient user interface development. In the second subset, some microcontrollers like ST, Raspberry pi, Arduino, etc. may be incorporated for MCU and MDA hub data processing. Further FPGA, CPLADs may also be used in MDA and processing units to enhance and to control/integrate the overall architecture. The proposed architecture utilizes a modular and flexible approach targeting multi-environment operations like under water operations, extreme heat and cold weather operations, considering different climatic conditions that may include humid, airly, raininy etc.

4.6 CONCLUSION

In this chapter, the author has proposed a wide range of hardware architecture for wearable systems incorporating the IoT as communication mainstream mechanism. The chapter details about the different communication protocols and their usability as per different applications and power range requirements. The power and range plays a major role in selection of the protocols for implementing efficient data exchange mechanism. Further the author discussed about the power harvesting advances which have become very handy for wearable solutions. The power requirements are also addressed with the help of new era circuit design techniques to have special power aware circuits for longer and better life and performance. The adiabatic circuit design techniques prove to be very effective in this era where longer battery life trading operating speed becomes handy for wearable systems. The foldable battery and circuits further assist in the advances of the system technologies. In this chapter, the author has provided a comprehensive system view, that should be considered for implementation of better and efficient wearable devices.

REFERENCES

1. M. A. Khan, M. T. Quasim, F. Algarni, and A. Alharthi, (2020) Internet of Things: On the Opportunities, Applications and Open Challenges in Saudi Arabia, in Proc. Int. Conf. on Advances in the Emerging Computing Technologies, Al Madinah Al Munawwarah, Saudi Arabia, pp. 1–5, https://doi.org/10.1109/AECT47998.2020.9194213
2. M. A. Khan and K. A. Abuhasel, (2021) Advanced Metameric Dimension Framework for Heterogeneous Industrial Internet of Things, Computational Intelligence, vol. 37, pp. 1367–1387, https://doi.org/10.1111/coin.12378

3. Dan Sung, What Is the Internet of Things? The Tech Revolution Explained Everything in Your Life, Connected to the Internet, https://www.wareable.com/features/what-is-the-internet-of-things-examples-definition, accessed on March 25, 2021.
4. Lionel Sujay Vailshery, (2021) IoT Connected Devices Worldwide 2030, Number of Internet of Things (IoT) Connected Devices Worldwide in 2018, 2025 and 2030, Technology & Telecommunications Consumer Electronics, https://www.statista.com/statistics/802690/worldwide-connected-devices-by-access-technology/#:~:text=Forecasts%20suggest%20that%20by%202030,from%20smartphones%20to%20kitchen%20appliances, accessed on February 2021.
5. N. S. Alghamdi and M. A. Khan, (2021) Energy-Efficient and Blockchain-Enabled Model for Internet of Things (IoT) in Smart Cities, Computers, Materials & Continua, vol. 66, no. 3, pp. 2509–2524.
6. Avnet, (2020) Wi-Fi 6 Is Here. Is It Time to Make a Move?, [Online]. https://www.avnet.com/wps/portal/apac/resources/article/wifi6-is-here/, accessed on April 5, 2021.
7. Avnet, (2020) How LoRaWAN and AI at the Edge Revolutionize the IoT, [Online]. https://www.avnet.com/wps/portal/apac/resources/article/how-lorawan-and-ai-at-the-edge-revolutionize-the-iot/, accessed on April 5, 2021.
8. Dima Kilani, Mohammad Alhawari, Baker Mohammad, Hani Saleh, and Mohammed Ismail, (2016, October) An Efficient Switched-Capacitor DC-DC Buck Converter for Self-Powered Wearable Electronics, Electrical and Computer Engineering Department, Khalifa Semiconductor Research Center (KSRC) Khalifa University of Science, Technology and Research, Abu Dhabi, IEEE Transactions on Circuits And Systems—I: Regular Papers, vol. 63, no 10, pp. 1557–1566.
9. Chris Smith, (2019) What Is Wearable Tech? Everything You Need to Know Explained, Article, https://www.wareable.com/wearable-tech/what-is-wearable-tech-753, accessed on March 2021.
10. Kiana Tehrani and Andrew Michael, (2014, March) Wearable Technology and Wearable Devices: Everything You Need to Know, Wearable Devices Magazine, WearableDevices.com, accessed on October 2017.
11. Ananth Dodabalapur, (2008) Flexible Hybrid Electronic Systems, University of Texas, https://slideplayer.com/slide/8721717/, accessed on March 2021.
12. M. A. Khan, (2020) An IoT Framework for Heart Disease Prediction Based on MDCNN Classifier, IEEE Access, vol. 8, pp. 34717–34727, doi: 10.1109/ACCESS.2020.2974687
13. A. Pantelopoulos and N. G. Bourbakis, (2010, Jan.) A Survey on Wearable Sensor-Based Systems for Health Monitoring and Prognosis, IEEE Transactions on Systems, Man, and Cybernetics, Part C (Applications and Reviews), vol. 40, no. 1, pp. 1–12, doi: 10.1109/TSMCC.2009.2032660
14. Kacey Deamer, (2016) Bendable Battery May Power Future Wearable Devices, Smartphones, https://www.livescience.com/56514-bendable-lithium-ion-battery.html, accessed on March 2021.
15. Muhammad Mahtab Alam and Elyes Ben Hamida, (2014) Surveying Wearable Human Assistive Technology for Life and Safety Critical Applications: Standards, Challenges and Opportunities Sensors, vol. 14, pp. 9153–9209, doi: 10.3390/s140509153, https://www.researchgate.net/publication/262607199_Surveying_Wearable_Human_Assistive_Technology_for_Life_and_Safety_Critical_Applications_Standards_Challenges_and_Opportunities/citations, accessed on February 2021.
16. Thang Viet Tran and Wan-Young Chung, (2016, December 15) High-Efficient Energy Harvester with Flexible Solar Panel for a Wearable Sensor Device, IEEE Sensors Journal, vol. 16, no. 24, pp. 9021–9028.
17. Farzanah Abdul Samad, Muhammad Faeyz Karim, Varghese Paulose, and Ling Chuen Ong, (2016, April 1) A Curved Electromagnetic Energy Harvesting System for Wearable Electronics, IEEE Sensors Journal, vol. 16, no. 7, pp. 1969–1974.

18. J. Carrillo and D. Marusiak, (2012) Energy Harvesting of Human Kinetic, Movement, Dept. Elect. Eng., California Polytechnic State University, San Luis Obispo, CA, Tech. Rep., 2012.
19. H. Liu, S. Gudla, F. A. Hassani, C. H. Heng, Y. Lian, and C. Lee, (2015, April) Investigation of the Nonlinear Electromagnetic Energy Harvesters from Hand Shaking, IEEE Sensors Journal, vol. 15, no. 4, pp. 2356–2364.
20. Longhan Xie and Mingjing Cai, (2014) Human Motion: Sustainable Power for Wearable Electronics, IEEE Pervasive Computing, vol. 13, no. 4, pp. 42–49.
21. Z. Zhou, B. Yao, R. Xing, L. Shu, and S. Bu, (2015) E-CARP: An Energy Efficient Routing Protocol for UWSNs in the Internet of Underwater Things, IEEE Sensors Journal, vol. 16, no. 11, pp. 4072–4082.
22. Grand View Research Wearable Sensors Market Size, Share & Trends Analysis Report by Sensor Type, by Device (Smart Watch, Fitness Band, Smart Glasses, Smart Fabric), by Vertical, by Region, and Segment Forecast, 2018–2025, https://www.grandviewresearch.com/industry-analysis/global-wearable-sensor-market, accessed on October 24, 2018.
23. J. Kathuria, M. A. Khan, A. Abraham, and A. Darwish, (2014) Low Power Techniques for Embedded FPGA Processors. In: M. Khan, S. Saeed, A. Darwish, and A. Abraham (eds) Embedded and Real Time System Development: A Software Engineering Perspective. Studies in Computational Intelligence, vol. 520. Springer, Berlin, Heidelberg. https://doi.org/10.1007/978-3-642-40888-5_11
24. Manoj Sharma, Asheesh Shah, and Arti Noor, (2016) Analysis and Evaluation of Adiabatic PFAL Inverter, International Journal of Control Theory and Applications, vol. 9, no. 23, pp. 299–309.
25. Manoj Sharma and Arti Noor, (2015) Reconfigurable CPLAG and Modified PFAL Adiabatic Logic Circuits, Advances in Electronics, Hindawi, vol. 2015, Article ID 202131, 10 pages, http://dx.doi.org/10.1155/2015/202131
26. Manoj Sharma and Arti Noor, (2014, March) Modified CPL Adiabatic Gated Logic – MCPLAG based DPET DFF with XOR, International Journal of Computer Applications (0975–8887), vol. 89, no. 19, pp. 35–41.
27. Manoj Sharma and Arti Noor, (2014, June) Reconfigurable CPL Adiabatic Gated Logic – RCPLAG based Universal NAND/NOR Gate, International Journal of Computer Applications (0975–8887), vol. 95, no. 26, pp. 27–32.
28. V. Leonov, (2010, November) Energy Harvesting for Self-Powered Wearable Devices, in Wearable Monitoring Systems, Springer, pp. 27–49.
29. Nadine Gergel-Hackett, Behrang Hamadani, Curt Richter, and David Gundlach, (2008) NIST USA, https://slideplayer.com/slide/15059816/, accessed on February 2021.
30. Ramesh Ramadoss, Simone Lee, Y. C. Lee, V. M. Bright, and K. C. Gupta, (2003, August) Fabrication, Assembly, and Testing of RF MEMS Capacitive Switches Using Flexible Printed Circuit Technology, IEEE Transactions on Advanced Packaging, vol. 26, no. 3, pp. 248–254.

5 Cache Memory Design for the Internet of Things

Reeya Agrawal and Neetu Faujdar
GLA University
Mathura, India

CONTENTS

5.1 Introduction 76
5.2 Problem Statement 78
5.3 Motivation and Overview 78
5.4 The Importance of Low Power VLSI Design 78
5.5 Contribution of the Chapter 79
5.6 Related Work / Literature Survey 79
5.7 Cache Memory Architecture 80
    5.7.1 Write Driver Circuit 80
    5.7.2 SRAMC Schematic and Working 81
    5.7.3 Sense Amplifiers 82
        5.7.3.1 Differential-Type Sense Amplifier 82
        5.7.3.2 Latch-Type Sense Amplifier 86
5.8 Methodology 87
    5.8.1 Power Reduction Sleep Transistor Technique (PRSTT) 88
    5.8.2 Power Reduction Dual Sleep Technique (PRDST) 88
    5.8.3 Power Reduction Sleepy Stack Technique (PRSST) 89
    5.8.4 Power Reduction Forced Stack Technique (PRFST) 89
5.9 Result and Discussion 89
    5.9.1 Simulation of Sense Amplifier 90
        5.9.1.1 Cache Memory Architecture with Voltage-Mode Differential Sense Amplifier (CMAVMDSA) 90
        5.9.1.2 Cache Memory Architecture with Current Mode Differential Sense Amplifier (CMACMDSA) 92
        5.9.1.3 Cache Memory Architecture with Charge Transfer Differential Sense Amplifier (CMACTDSA) 94
        5.9.1.4 Cache Memory Architecture with Voltage Latch-Type Sense Amplifier (CMAVLTSA) 94
        5.9.1.5 Cache Memory Architecture with Current Latch-Type Sense Amplifier (CMACLTSA) 96
5.10 Comparison Table 98
5.11 Conclusion and Future Scope 101
References 102

DOI: 10.1201/9781003122357-7

5.1 INTRODUCTION

The demand for mobile devices and battery-operated embedded systems is growing with greater breadth as the very large-scale integrated circuit (VLSI) industries grow. Cache memory is a central part of memory that plays a key role in data execution, occupying 60% to 70% of the chip region [1]. As chip consumption increases rapidly, microprocessor velocity decreases. One million transistors also increase and degrades the efficiency of single-chip failure rates, so the industry is working to build a low-speed and low-power memory circuit that keeps the development of the VLSI system informed. In this article, the emphasis is on the sense amplifier. In current high-performance microprocessors, more than half of the transistors are for cache memories, and in the future, this proportion is projected to increase [2]. SRAMC is usually the option for built-in stock because it is robust in such chips in a noisy environment. The design of low-power, high-performance processors, therefore, received considerable attention. The device can use necessary memory cells by integrating them into SRAMCs that are the right size for system requirements. In an area, speed and power lead to improvements. In all SRAMC memory blocks, SA is an important element that responds to high frequency. Memory time for access and power consumption are calculated primarily by the configuration of SA. SA is one of the most important peripheral circuits in memory systems [3, 4]. A SA is a power-operated circuit that reduces the time it takes for a signal to travel from a cell to a logic circuit on the memory cell periphery and transforms arbitrary logical levels of peripheral Boolean circuits to digital logic levels [5]. Their output significantly influences both memory access time and the total dissipation of memory capacity. Complementary metal-oxide-semiconductor (CMOS) memories, as with other integrated circuits (ICs), are needed to increase speed, increase power, and keep dissipation low. When it comes to SA memory design, these goals are quite contradictory. Usually, with increased memory space, the parasite space of the bit-line is also increased. This bit-line has increased steadily with more energy-hungry memories [6].

In the emerging trend of a VLSI industry, applications of semiconductors have been increased in various fields such as agriculture, medical industrial settings, and the Internet of Things (IoT) [7–9]. The functional parameters are increasing day by day on an IC, which is nothing but an increase in power consumption [10]. CMOS scaling nowadays ranges from 10 μm in 1971 to 5 ηm now. In today's technological world, low power dissipation devices are required. Since portable handheld devices have a low number of power plugs in their surroundings, i.e., they need a high battery backup system that consumes less energy during sleep mode and active mode [11].

As shown in Figure 5.1, a typical IoT sensor includes four main devices: a sensor, a microcontroller (MCU), a wireless computer (Bluetooth, Wi-Fi, LoRa), and a static random-access memory cell (SRAMC). The devices communicate with each other over a serial bus, such as the serial peripheral interface (SPI) or inter IC communication (I2C) protocol, since sensor nodes operate at low data rates and low power [12]. To negotiate all transactions between devices, the MCU serves as the master controller for the IoT node. The wireless system must send sensor data to the nearest gateway and also receive control instructions from the gateway. The sensor

Cache Memory Design for the Internet of Things

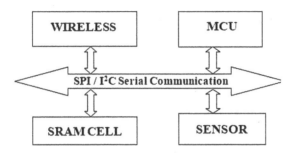

FIGURE 5.1 Block diagram of a typical IoT sensor node.

tests environmental parameters (temperature, humidity, etc.) and transmits the data digitally through the serial communication bus of the SPI/I2C [13]. When a gateway is not available for communication, the SRAMC is used to store local data. The innovation of this project is to generate and monitor all the signals needed for the SRAMC without using internal clocks, instead of deriving all control signals from the SPI signals themselves, making it a simple and scalable circuit that can be made with slower access time to consume very little power [14].

Cache memory is a very high-speed memory that consumes very little power. Cache memory is used to speed-up for synchronizing with a high-speed central processing unit (CPU). It is used as a CPU register because of its economical behavior even though it is costlier than main memory [15]. Cache memory serves as a buffer between the CPU and random access memory (RAM), as the memory is extremely fast. It has data and instructions holding capacity so that it is easily and instantly accessible to the CPU. SRAMC reduces the average time for data access and has a holding capacity of data than DRAM [16].

VLSI chip is the most crucial aspect of SRAMC-based cache. Peripheral circuit and memory performance can undesirably influence overall system speed and power. The most important problem in SRAMC cache design is the delay in the reading operation. After the latching process, the current flow stops automatically. Thus, static power is not dissipated in SA. CMOS memory performance is affected by the time and total memory power usage of SA, due to this SA is a crucial component of CMOS memory. The design of an ultra-low-voltage SRAMC is important for decreasing the total system's power dissipation in applications like biomedicine, wireless sensors, and implantable electronics. Owing to their compact and basic form, the traditional SRAMC is widely used [17]. However, the opposing read and writes specification constrains the minimum operating voltage (V_{min}).

Cross-coupled types of CMOS inverters are simple SAs. The output lines function at the same time as the input lines. There is a lot of dissipation for both delay and power. The memory cells or the whole SRAMC blocks are of particular importance in this work, although the SRAMC SAs do not attract broad attention. In today's world, technology needs low power dissipation devices due to portable handheld devices' low number of power plugs in their surroundings. They need a high-battery backup device that consumes less energy during the non-working process as well as during the working process [18].

Sense amplifiers are used primarily to read SRAMC content and dynamic random-access memory (DRAM). Their conception means an acceptable spectrum of noise and high-quality information that reflects the material of a specific memory cell. They are very sensitive to noise. Fast sense amplifiers are critical for many circuits to achieve low latency, with bit-line reading in memories being the most common domain [19].

5.2 PROBLEM STATEMENT

One of the key problems when developing a VLSI device is power dissipation. Dynamic power was the single biggest problem up to a certain time; but, as the scale of the technology function shrinks, static power as dynamic power has become an important issue. To save leakage power consumption, a well-known previous approach called the sleep transistor technique cuts off V_{DD} or ground connections of transistors. However, if transistors are allowed to float, a device may have to wait a long time to recover the missing state consistently and therefore may suffer severely degraded performance. For a device that needs a quick response even though in an inactive state, maintaining a state is therefore essential. But it does have a deferred penalty and contrasts the field requirement with other procedures. Once again, the sleepy keeper process has excellent pace requirements, but more static and dynamic capacity than the sleepy stack is needed. Our aim is the trade-off between these restrictions and hence recommend new approaches with the least possible region and delay trade-offs that reduce both leakage and dynamic capacity.

5.3 MOTIVATION AND OVERVIEW

Modern digital systems require high-speed memories for storing and retrieving large amounts of data. SRAMC is widely used for all memories because of its high speed and low power consumption. According to the 2002 ITRS (International Technology Roadmap for Semiconductors), memory will occupy 90% of the chip by 2017.The operating speed is increased; the chip size also increases, so with the increase in chip size, the power consumption by the circuit becomes very important [20].

Increasingly, the speed of the VLSI chip is constrained by the signal latency of long interconnection lines. When using CMDSA rather than VMDSA signal transport methods, significant speed and power changes are feasible. The memory cell size can also be minimized with the current mode sensing. Since the majority of memory-related activities are read operations, the memory's net power dissipation is greatly reduced. SAs dissipate a significant amount of short circuit power, a significant amount of dynamic power is saved by the cell sequence [21].

5.4 THE IMPORTANCE OF LOW POWER VLSI DESIGN

In VLSI, the reduction in feature size and a related rise in chip density and operating frequency continues to be a major concern for power consumption. It causes overheating, which reduces the chip lifetime and degrades the performance of the portable systems. The need for the optimization of power consumption in a chip is increased because of the need for portable communication devices and computing systems. As going toward the integration in VLSI circuits, the importance of low power is increasing day by day. Many factors affect the power consumption in the circuits [22].

- C_{BL} increases as the number of memory cells per bit-line increases, while R_{BL} increases as the bit-line length increases.
- Current records are being lowered when more memory is incorporated on a single chip for heavy capacity. This results in a lower voltage swing on the bit-line, as well as a rise in C_{BL}.
- Low supply voltage results in a lower margin of noise.

5.5 CONTRIBUTION OF THE CHAPTER

This research aims to develop a single-bit cache memory architecture with low power consumption. This rule guarantees that the value stored during the write operation is exchanged in the bit-cell that can be used in memory blocks. VMDSA, CMDSA, CTDSA are three types of differential-type sense amplifiers and VLTSA, CLTSA are two types of latch-type sense amplifiers. Various cache memory architecture blocks are introduced in addition to multiple types of power reduction methods, such as sleep transistor technique, forced stack technique, sleep stack technique, and dual sleep technique are applied over different blocks of cache memory architecture such as SRAMC and SAs.

5.6 RELATED WORK / LITERATURE SURVEY

Zhao et al. [23] proposed a new proposal which was developed for the use of carbon nanotube field transistors in a ternary SRAMC sensory amplifier (CNFET). The threshold voltage is regulated by the CNFET chirality to grasp the ternary logic. Compared with a ternary DRAM and the ternary SRAMC without SA simulation is conducted on HSPICE at 0.9 V.

Pahuja et al. [24] describe three distinct power reduction schemes in the SRAMC and charge-recycling, power gating, and low-energy architecture were incorporated in this article. This documentation contains 45 nm technology, showing that the average power decreased by 1/10th in amplifier design and also a decrease in read and write cycle latency of 64-bit SRAMC array architecture.

Kyung et al. [25] System-on-chip (SOC) models have been noted for providing a set of multi-port memory IP blocks supporting multiple simultaneous operations in the same memory bank to increase performance. Conventional 2-read/write 8T dual-port SRAMC (2RW) has reading and typing troubles when both words are activated in a single row at the same time. 1-read, 1-write 8T dual-port cells (1R1W) mitigate reading disruption by preventing load sharing with internal storage nodes when the read word line is activated (RDWL).

YoungBac et al. [26] describe that leakage power consumption in deep-submicron technology has become a major concern in VLSI circuits, particularly in SRAMCs that are used to build the system-on-chip cache. A low power 8T SRAMC based on Carbon Nanotubes Field Effect Transistors (CNFET) is proposed to improvise the issue of leakage power in this paper.

Stefan *et al.* [27] describe that in particular because of biased temperature instability, ICs are shown to be more susceptible to transistor ageing with chronic downgrades in CMOS technologies. This paper offers a benchmarking of the BTI effect in 3 modes at various voltages and supply temperatures: low power (LP), middle power/output (MP), and high power (HP). The SD of the three prototypes is estimated on the basis of a 45 nm technology for different working loads.

Bingyan *et al.* [28] It means that there are higher deviations and incompatibilities in the transistors that lead to higher offset voltages with the CMOS technology continuing to decrease. Speed can be improved by broad offset voltage and increased dynamic power use via a reading process, degrading the correct decision rate of sensing and running speed. As a result of transistor threshold-voltage error, offset voltage is thus the most important metric for SRAMC sensory amplifiers (SAs). The authors proposed the digitizing multiple body bias offset cancelling strategy (DMBB). The SA transistors' threshold voltage mismatch is balanced by adjusting the body-biased voltage in this system in a digital and continuous manner.

Mottaqiallah *et al.* [29] Researchers report that a lot of work has been published in SRAMC on influences from BTI, but many papers concentrate only on the collection of memory cells. SRAMC also includes peripheral circuits such as address decoders, sense enhancements, and so on. This paper discussed the cumulative impact of BTI on the sensory amplifier of various technology nodes and changes in voltage temperature. The assessment measure is tested for a range of tasks. This paper provides a quantification of BTI's better influence on all technology scaling parameters in comparison to past work.

ByungKy *et al.* [30] Due to increased procedure variances and a reduction in supply tension, the read yield was discovered to be deteriorating. The author proposes a LOC-SA (latch offset sensory amplifier) that cancels the latch offset function, leading to considerable voltage growth time decreases and sensing speed increases.

Hanwool *et al.* [31] They describe the increase in efficiency and the minimization of power consumption by using the same pull-up PMOS transistor in order to detect and preload the bit-line. The guard also employs a bit-line leak compensator and has a minimal operating voltage.

5.7 CACHE MEMORY ARCHITECTURE

It is important to build cache memory architecture (CMA) as shown in Figure 5.2 so that a non-destructive read and a secure writing process are feasible. On SRAMC transistor sizing, these two requirements place conflicting requirements [32, 33].

5.7.1 WRITE DRIVER CIRCUIT

Figure 5.3 shows the schematic of WDC. Each of the bit-lines in the SRAMC write driver circuit is quickly discharged from pre-charge stages to below the SRAMC write margin.

Cache Memory Design for the Internet of Things

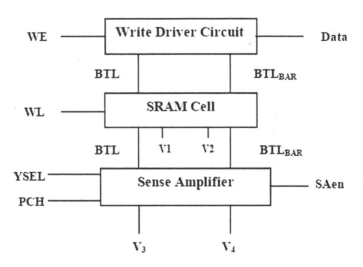

FIGURE 5.2 Cache memory architecture block diagram.

The write enable (WE) signal usually activates the WDC, which uses full-swing discharge to drive the bit-line from the pre-charge level to the ground. Five P_{MOS} (PM1, PM2, PM3, PM4, and PM5) as well as five N_{MOS} (NM1, NM2, NM3, NM4, and NM5) are used by WDC. When allowed by WE, the input data causes one of the transistors to become PM1 or NM1 through inverters, and a strong 0 is applied by discharging BTL and BTL_{BAR} from the pre-charge level to ground level [34].

5.7.2 SRAMC Schematic and Working

For a variety of reasons, design considerations of SRAMC are relevant. Firstly, the architecture of the SRAMC is important for the safe and robust functioning of the SRAMC. Secondly, the SRAMC designers are inspired to increase the packing density thanks to their continuous effort to increase the storage capacity

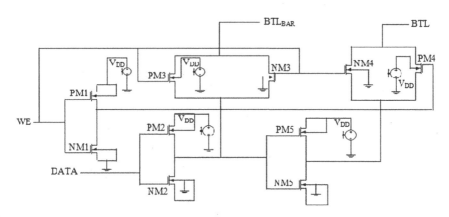

FIGURE 5.3 Write driver circuit schematic.

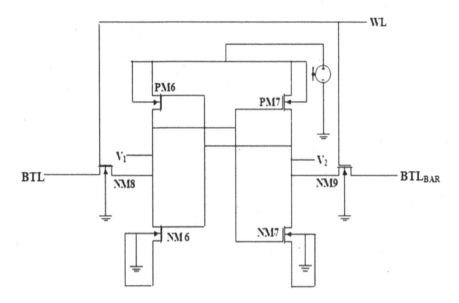

FIGURE 5.4 SRAMC schematic.

of chips. The SRAMC, therefore, needs to be as small and as stable, quick, powerful, and yield constraints as possible. The word line defines modes of operation, when all transistors are removed and cells are separated [35]. The word line pulls high (WL = 1) for reading and writes operations that allow access to access transistors (NM8 and NM9) to be enabled. A typical SRAMC uses two cross-connected inverters which create a transistor lock and access. In reading and write operations, access transistors (NM8 and NM9) allow access to the cell and provide cell isolation during the non-accessible state as shown in Figure 5.4. The architecture adopted for a cell should obey two basic rules for the proper functioning of SRAMC [36].

5.7.3 Sense Amplifiers

SA is an important memory design aspect. Because of the range of SAs in semiconductor memories and the effects they have on the final requirements, they become a different circuit class. Sensing operation must be non-destructive, provided that SRAMCs need no data to refresh circuits after sensing [37]. Table 5.1 shows difference between differential-type sense amplifier and latch-type sense amplifier.

5.7.3.1 Differential-Type Sense Amplifier

All the elements necessary for differential sensing are included in the basic metal oxide semiconductor (MOS) sense amplifier circuit. An amplifier is capable of rejecting popular noise and of amplifying the real difference between the indicators. Since the operating rate is very sluggish, the main difference between the amplifiers is not in memories due to significant power dispersion and an inherently high offset [38].

TABLE 5.1
Difference between Differential-Type and Latch-Type Sense Amplifier

Differential-Type Sense Amplifier	Latch-Type Sense Amplifier
Less access time or high delay	Fast access time or less delay
Due to more delay than latch type, it consumes less power than latch type	Power consumption is high
As it has pre-charge circuitry in its circuit, it gives high voltage swing	It has poor voltage swing.
The area is large, making it complicated to design.	Whereas its area is small, so easy to design

5.7.3.1.1 Voltage-Mode Differential Sense Amplifier

A voltage mode differential sense amplifier detects the differential voltage on the bit-lines and produces a full rail output. In Figure 5.5 the circuit is shown. When the word line is pulled high (WL = 1), BTL and BTL_{BAR} pair is used to produce a voltage differential. The sense amplifier is pulled high (SA_{EN}), which causes the cross-coupled inverter to enter into a positive feedback loop to convert its differentials to maximum rail output, which is triggered when appropriate differential depends on technology and circuit.

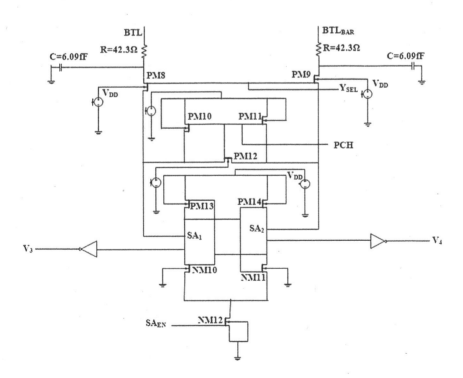

FIGURE 5.5 Voltage-mode differential sense amplifier schematic.

The sense amplifier output node connecting with the bit-line is down to zero while the other output SA_1 stays elevated with a lower voltage, e.g., SA_2. N_{MOS} such as NM10 and NM11 go into saturation when the sense amplifier is activated. The complete V_{DD} input NM11 device has a higher voltage than NM10 with a lower voltage of its V_{gs}. This positive feedback loop continues until the voltage SA_2 is lower enough for N_{MOS} device NM12 to go in the linear region and switch on the other inverter's P_{MOS} device PM13 and push the output extremely [39].

5.7.3.1.2 Current-Mode Differential Sense Amplifier

Figure 5.6 shows the circuit diagram of CMDSA. By sensing bit-cell current directly, CMDSA works. It does not rely on a different voltage that develops across the bit-line [40]. The lower bit-line can be clamped at a higher voltage as is necessary with an amplifier voltage sense. This can minimize bit-line power pre-charge. Two components of the current mode sense amplifier are a transmitting circuit with unit transmission characteristics and a sensing circuit that senses the differential current. The current mode sense amplifier functions in two steps: pre-charge and evaluation. The bit-lines are pre-loaded by a pre-charge circuit which attaches to the bit-lines during the pre-loading stage. The SA_1 and SA_2 sense amplifier output nodes are pre-loaded by P_{MOS} devices such as PM12 and PM13. The PM12 and PM17 devices are ON during pre-loading, which equalizes the sensing circuit inputs and outputs. Devices (PM14, PM15, NM10, and NM11) evaluate the operating stream of the sensor boost. The pre-load and equalizer PM12, PM12, PM16, and PM17 are switched OFF at the end of the pre-charge

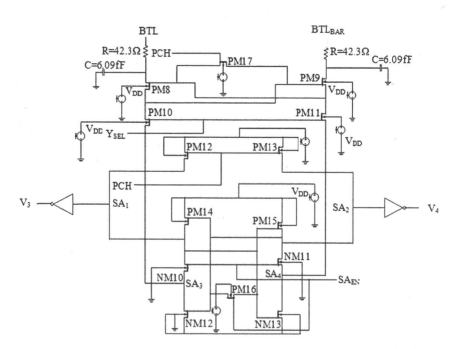

FIGURE 5.6 Current-mode differential sense amplifier schematic.

process. Y_{SEL} is pulled low ($Y_{SEL} = 0$) and the SA_{EN} is pulled high ($SA_{EN} = 1$) during the evaluation phase. A high benefit positive feedback amplifier then forms the cross-connected inverters made of devices PM14, PM15, NM10, and NM11. Because of the positive feedback, either NM10 or NM11 are impeded to the NM10 or NM11 source terminal, resulting in NM10 and NM11 beginning to source a portion of the difference. The delay in sensing is relatively insensitive to bit-line capacitance because the operation does not rely on differential voltage development along the bit-lines. The power nodes are not connected to high-capacity bit-lines, and they can react quickly, contrary to the sense amplifier. On bit-lines the CMDSA will swing lower voltage [41].

5.7.3.1.3 Charge Transfer Differential Sense Amplifier

By using charging redistribution from high-capacity bit-lines to low-capacity amplifier output nodes, the charge transfer differential sense amplifier operates. The effect is higher speed and lower energy consumption since the bit-lines have a low voltage swing [42], whereas Figure 5.7 shows CTDSA schematic.

As seen in the following equation, the voltage produced across the first element and its capabilities would equalize the voltage product across the second element and its voltage potential to a set of relationships of two capacitive elements in a charging device: $V_{SMALL} C_{SMALL} = Q = V_{LARGE} C_{LARGE}$.

Therefore, the voltage increase is achieved because a small voltage change over the large capacitive element causes a larger voltage change across the smaller capacitive element [43].

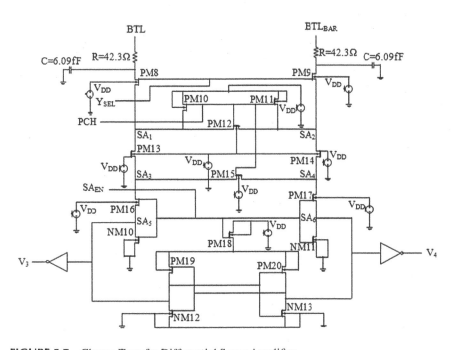

FIGURE 5.7 Charge-Transfer Differential Sense Amplifier.

5.7.3.2 Latch-Type Sense Amplifier

The sensing begins with pre-loading and equalizing the latch-type SA in the high-gain metastable region. Since the input is not segregated from the outputs in a latch-type SA, the insulation transistors are necessary for isolation of the latch-type SA of the bit-lines and preventing the complete unloading of a 0 bit-line that costs additional power and time [44, 45].

5.7.3.2.1 Voltage Latch-Type Sense Amplifier

Figure 5.8 shows the voltage latch-type sense amplifier schematics made in this work. The internal nodes of this design are pre-charged via the bit-lines. NM12 is OFF, and PM8 and PM9 pass transistors are ON when the word line is pulled high and before the sense amplifier signal.

As the differential increases on the bit-lines, the random bit on the sense amplifier's internal nodes of the sense amplifier has a suitable voltage difference. The PM10, NM10, PM11, and NM11 cross-linked inverters amplify the differential voltage to its maximum swing output when the SA_{EN} sense amplifier signal is asserted [46].

5.7.3.2.2 Current Latch-Type Sense Amplifier

A circuit of considerable significance in the design of cache memory is the sense amplifier. During the reading operation, one of the bit-line releases while the other bit-line stays at supply voltage as shown in Figure 5.9.

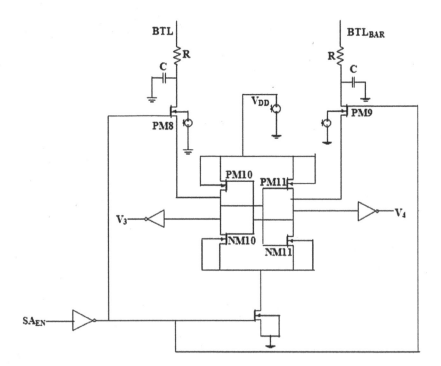

FIGURE 5.8 Voltage latch-type sense amplifier schematic.

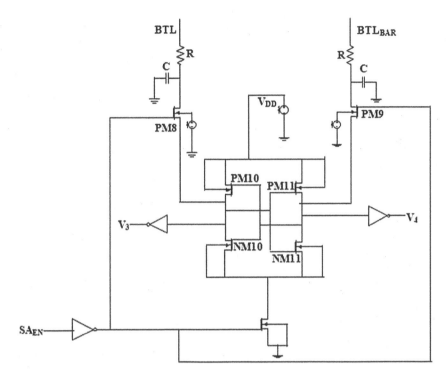

FIGURE 5.9 Current latch-type sense amplifier schematic.

The slow discharge is small, due to the ability of large bit-line, and the bit cells access the transistor. To achieve this, a tiny difference is amplified by the sense amplifier at digital levels between the values of the bit-line voltages. Figure 5.8 indicates the current latch sense amplifier schematic. The circuit operation is as follows [47]. The differential voltage is passed on bit-lines to SA_3 and SA_4 CLSA inputs. SA_{EN} is pulled high if both SA_1 and SA_2 begin discharge at high outputs. These results in a higher NM_{12} power compared to NM_{13} due to its higher V_{gs}. This makes it possible to discharge the V_3 output faster than the V_4. When the sense amplifier signal is low ($SA_{EN} = 0$) enough to power ON P_{MOS} device PM_{16}, the powerful positive feedback loop is triggered, which causes SA_2 to be recharged and its output isolated from its inputs.

5.8 METHODOLOGY

The fundamental operational approach for leakage power reduction methods is covered in this section, such as the sleep transistor technique, the dual sleep technique, the forced stack technique, and the sleep stack technique, which have been used to evaluate various parameters in the SRAMC circuit [48]. These techniques for leakage reduction can be grouped into two categories:

1. State saving
2. State destructive

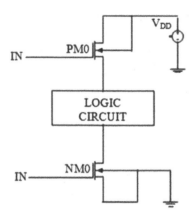

FIGURE 5.10 PRSTT schematic.

5.8.1 Power Reduction Sleep Transistor Technique (PRSTT)

The most common power reduction technique is sleep transistor technique. Figure 5.10 shows the schematic of sleep transistor technique. In sleep transistor technique (i) between V_{DD} and the pull-up network in circuit PMOS is placed, whereas (ii) between pull-down network and GND in circuit NMOS is placed [49].

5.8.2 Power Reduction Dual Sleep Technique (PRDST)

In this technique, four transistors, two PMOS (PM0 and PM1) and two NMOS (NM0 and NM1), are added, the area specifications for this technique are maximized. The dual sleep technique has taken advantage of the two extra pull-up and pull-down transistors in sleep mode in either an OFF/ON state. As seen in Figure 5.11, fewer transistors are needed for a specific logic circuit [50].

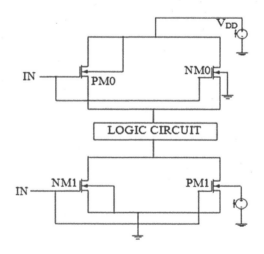

FIGURE 5.11 PRDST schematic.

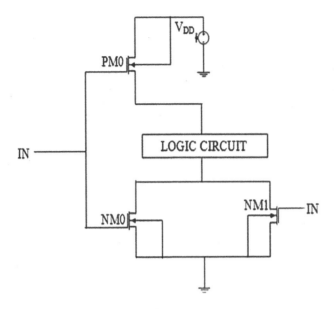

FIGURE 5.12 PRSST schematic.

5.8.3 POWER REDUCTION SLEEPY STACK TECHNIQUE (PRSST)

The stack method, which forces a stack effect by breaking down an existing transistor into two half-sized transistors, is another technique for power reduction as shown in Figure 5.12. The induced reverse bias between the two transistors results in a sub-threshold reduction of leakage current when the two are switched OFF together. Split transistors, however, dramatically increase delay and may reduce the utility of the approach [51].

5.8.4 POWER REDUCTION FORCED STACK TECHNIQUE (PRFST)

In the forced stack technique instead of using voltage supply PM0 is used and in place of ground NM0 is used in the logic circuit as shown in Figure 5.13. In this technique, both MOS have the same input. In this technique when PM0 is in the active region, NM0 is in the cut-off region [52]. Due to this, the circuit doesn't have a power supply all the time and this helps to consume less power.

5.9 RESULT AND DISCUSSION

In this chapter, the discussion on methodology and results has been completed. As the feature size of technology decreases, the threshold voltage of MOS also decreases. Because of the scaled down threshold voltage, the MOS transistor is turned on at a lower voltage, so the delay of the circuit decreases, and the power dissipation as the supply voltage decreases.

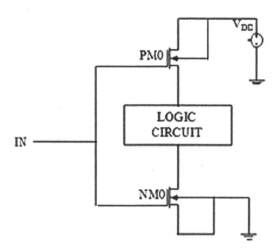

FIGURE 5.13 PRFST schematic.

5.9.1 SIMULATION OF SENSE AMPLIFIER

Sense amplifiers are used in reading circuitry and amplify the signal. There are different types of sense amplifier circuits. Here, the simulation of five types of sense amplifiers is categorized into two different categories, i.e., differential-type and latch-type sense amplifiers.

5.9.1.1 Cache Memory Architecture with Voltage-Mode Differential Sense Amplifier (CMAVMDSA)

The schematic of CMAVMDSA is shown in Figure 5.14. In the schematic, internal node voltages out and outb are initialized with the logic 0 and 1, respectively.

Both the bit-lines are pre-charged by the pre-charge circuitry applied to the signal PCH = 0 V. Sense amplifier is enabled by signal SA_{en} = 1.

Figure 5.15 is the output waveform of WDC, where WE and DATA are inputs and BTL and BTL_{BAR} are outputs as described in the following four cases. Case 1: when WE are low and data is low, BTL = V_{DD} and BTL_{BAR} = V_{DD}; Case 2: when WE are high and data is low, BTL = 0V and BTL_{BAR} = high; Case 3: when data is high and WE are low, BTL = BTL_{BAR} = V_{DD}/2; and Case 4: when data is high and WE are high, BTL = V_{DD} and BTL_{BAR} = 0V. Figure 5.16 shows the output waveform of SRAMC, in the output waveform hold operation and write operation both are shown as when WL is high, write operation is done and when WL is low, the memory cell holds the stored data and logic 1 is stored in the cell from the difference of V_1 and V_2.

Figure 5.17 is the output waveform of CMAVMDSA. From this, Y_{SEL} = 0, the pre-charge circuits remain high to pre-charge the bit-lines (i.e., P_{ch} = V_{DD}), the data on the bit-lines is passed through, the WL of SRAMC remains high (i.e., WL = $_{VDD}$) for reading operation, i.e., the access transistors (NM8 and NM9), BTL = 0V and BTL_{BAR} = V_{DD}, the voltage difference is sensed by the SA, and the stored data has been sensed.

Cache Memory Design for the Internet of Things 91

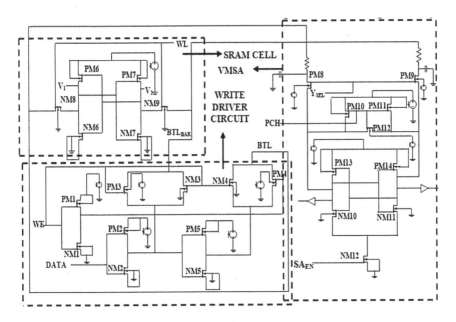

FIGURE 5.14 Cache memory architecture with voltage mode differential sense amplifier schematic.

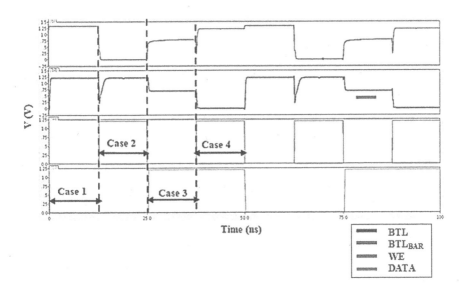

FIGURE 5.15 The output waveform of a write driver circuit.

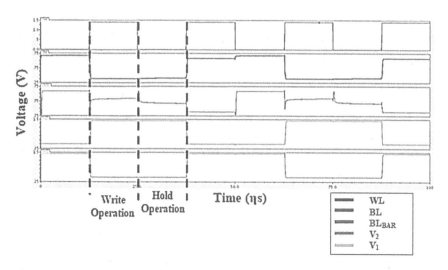

FIGURE 5.16 The SRAMC output waveform.

5.9.1.2 Cache Memory Architecture with Current Mode Differential Sense Amplifier (CMACMDSA)

The schematic of CMACMDSA is shown in Figure 5.18. Signal activated the sense amplifier, when SA_{en} is high. A pre-charge signal (PCH) is used to pre-charge the bit-lines to equalize the potential on both bit-lines. In the schematic, the internal node voltages V_1 and V_2 are initialized with logic 0 and logic 1, respectively.

The output waveform of CMACMDSA is shown in Figure 5.19. YSEL = 0V, pre-charge circuits keep high to pre-charge the bit-lines (i.e., Pch = VDD), the current on

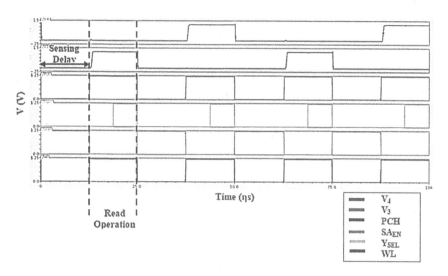

FIGURE 5.17 Cache memory architecture with voltage mode differential sense amplifier output waveform.

Cache Memory Design for the Internet of Things

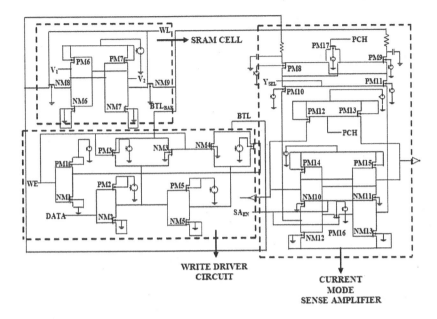

FIGURE 5.18 Schematic of a cache memory architecture with a current mode differential sense amplifier.

the bit-lines is passed through and stored at nodes as transistors are in active mode. WL of SRAMC keeps high (i.e., WL = VDD) for reading operation, i.e., switched ON the access transistors (NM8 this time, the sense amplifier senses the difference between BTL and BTL_{BAR} and the stored data has been sensed through the sense

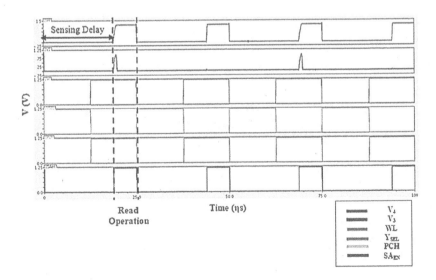

FIGURE 5.19 Cache memory architecture having current mode differential sense amplifier output waveform.

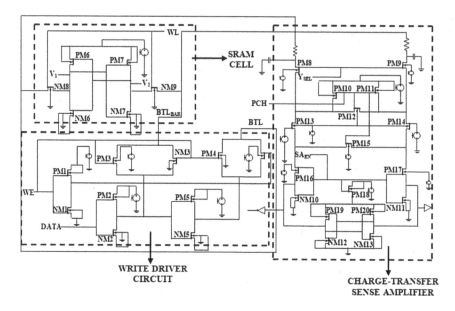

FIGURE 5.20 Cache memory architecture having charge transfer differential sense amplifier schematic.

amplifier and is shown at V_4, i.e., logic 1 is stored in the memory and hence sensed by the sense amplifier.

5.9.1.3 Cache Memory Architecture with Charge Transfer Differential Sense Amplifier (CMACTDSA)

Figure 5.20 shows the schematic of CMACTDSA. Charge relocation from high capacitance bit-lines to low capacitance bit-lines is used in this system.

Figure 5.21 shows the output waveform of CMACTDSA, $Y_{SEL} = 0$, pre-charge circuits kept high to pre-charge the bit-lines (i.e., $P_{ch} = V_{DD}$), data on the bit-lines is passed from SRAMC to sense amplifier, WL of SRAMC kept high (i.e., WL = V_{DD}) for reading operation, i.e., switched ON the access transistors (NM8 and NM9), BTL = 0V and $BTL_{BAR} = V_{DD}$, i.e., voltage difference is V_{DD} at the output, $SA_{EN} = 0V$ (for half positive cycle of WL). At this time, the sense amplifier senses the difference between BTL and BTL_{BAR}, and the stored data has been sensed by the sense amplifier and is shown at V_4, i.e., logic 1 is stored in the memory and hence sensed by the SA.

5.9.1.4 Cache Memory Architecture with Voltage Latch-Type Sense Amplifier (CMAVLTSA)

Figure 5.22 shows the schematic of CMAVLTSA. Bit-lines are used to pre-charge the internal nodes of this architecture. The input bit-lines establish a voltage differential on the circuit's internal nodes, which the circuit architecture relies on directly.

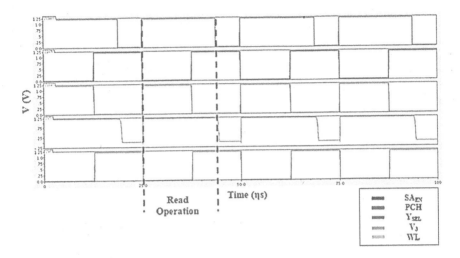

FIGURE 5.21 Cache memory architecture having charge transfer differential sense amplifier output waveform.

Figure 5.23 shows the output waveform of CMAVLTSA. $Y_{SEL} = 0V$, pre-charge circuits keep high to pre-charge the bit-lines (i.e., $P_{ch} = V_{DD}$), the current on the bit-lines is passed through and is stored at nodes as transistors are in active mode, WL of SRAMC keeps high (i.e., WL = V_{DD}) for reading operation, i.e., switched ON the access transistors (NM8 and NM9), BTL = 0V, $BTL_{BAR} = V_{DD}$ (for half the positive cycle of WL). At this time, the sense amplifier senses the difference between BTL and BTL_{BAR}, and the stored data has been sensed through the sense amplifier and is shown at V_4, i.e., logic 1 is stored in the memory and hence sensed by the SA.

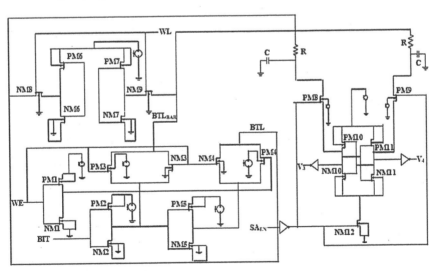

FIGURE 5.22 A voltage latch-type sense amplifier schematic is used in a cache memory architecture.

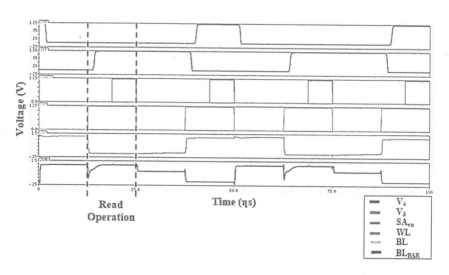

FIGURE 5.23 Cache memory architecture having voltage latch-type sense amplifier output waveform.

5.9.1.5 Cache Memory Architecture with Current Latch-Type Sense Amplifier (CMACLTSA)

Figure 5.24 shows the schematic of CMACLTSA. In the schematic, internal node voltages V_1 and V_2 are initialized with logic 0 and logic 1, respectively.

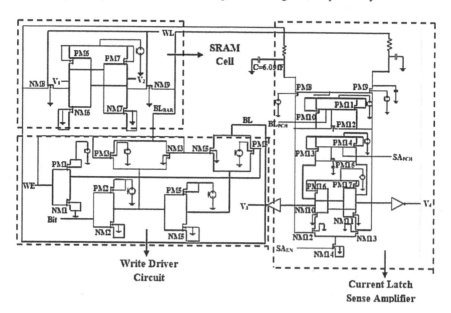

FIGURE 5.24 Cache memory architecture having a current latch-type sense amplifier schematic.

Cache Memory Design for the Internet of Things

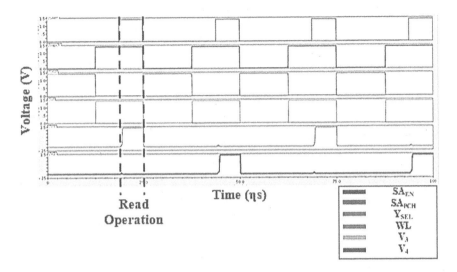

FIGURE 5.25 Cache memory architecture having current latch-type sense amplifier output waveform.

Figure 5.25 describes the read operation of CMACLTSA when both SA_{EN} and WL are pulled high. During that time, only SA senses the data from the SRAMC at bit-lines and gives output at V_3 and V_4.

Figure 5.26 shows the process corner simulation of output V_3 of SA. Figure 5.27 shows the Monte Carlo simulations of $V_{TH\_SA_{EN}}$, on which the SA of a cache memory architecture depends.

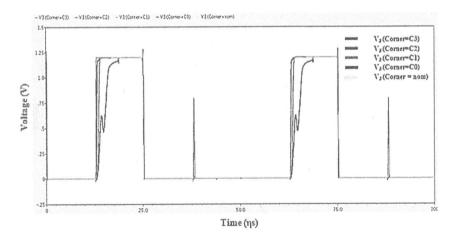

FIGURE 5.26 Process corner simulation cache memory architecture.

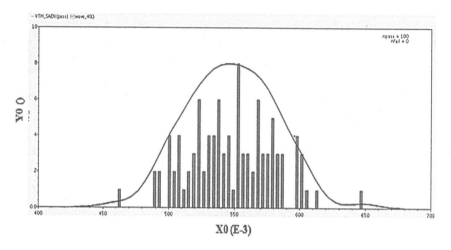

FIGURE 5.27 Monte Carlo simulation cache memory architecture.

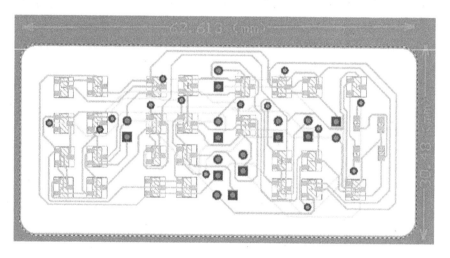

FIGURE 5.28 Chip design of cache memory architecture having voltage mode differential sense amplifier.

Figure 5.28 shows the chip design of voltage mode differential sense amplifier and it uses 62.613×30.48 mm^2 area which is less than other circuit areas as shown in Figures 5.29 and 5.30.

5.10 COMPARISON TABLE

The power consumption of all the circuits has been analyzed in this section. Apart from its number of transistors and sensing delay, each circuit has been compared at different values of R as shown in Table 5.2. It compares the power consumption, sensing delay, and number of transistors of the cache memory architecture using different types of SA with a resistance of 42.3Ω.

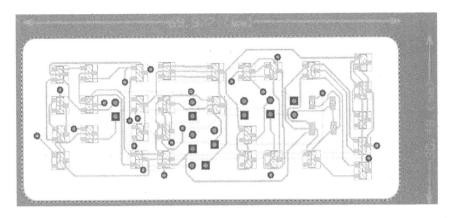

FIGURE 5.29 Chip design of cache memory architecture having current mode differential sense amplifier.

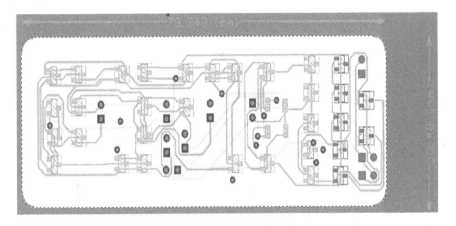

FIGURE 5.30 Chip design of cache memory architecture having charge transfer differential sense amplifier.

TABLE 5.2
Power Consumption of Single-Bit Cache Memory Architecture When $V_{DD} = 1.2$ V, C = 6.09 fF, and R = 42.3Ω

S. No.	Parameters Architecture	Total Power Consumption	Sensing Delay	Number of Transistors
1.	SBSVMDSA	13.16 µW	13.51 ηs	30
2.	SBSCMDSA	16.44 µW	18.81 ηs	33
3.	SBSCTDSA	44.63 µW	18.95 ηs	37
4.	SBSVLTSA	36.57 µW	13.50 ηs	29
5.	SBSCLTSA	26.78 µW	18.68 ηs	35

TABLE 5.3
Power Consumption of Cache Memory Architecture When $V_{DD} = 1.2$ V, $C = 6.09$ fF, and $R = 42.3$ KΩ

S. No.	Parameters Architecture	Total Power Consumption	Sensing Delay	No. of Transistors
1.	SBSVMDSA	11.34 µW	13.51 ηs	30
2.	SBSCMDSA	18.81 µW	18.81 ηs	33
3.	SBSCTDSA	33.63 µW	18.95 ηs	37
4.	SBSVLTSA	14.32 µW	13.50 ηs	29
5.	SBSCLTSA	73.92 µW	18.68 ηs	35

Table 5.3 concludes that the cache memory architecture having the VMDSA consumes the least power among all the SAs. Due to this, PRT has been applied over a SRAMC VMDSA in a cache memory architecture, but there is an increment in the number of transistors in the design which depicts that area increases with decreasing power consumption.

Table 5.4 describes the power consumption of cache memory architecture having VMDSA and to optimize power, PRT are applied over VMDSA and there is an increment in the number of the transistor which depicts that on optimizing power the area increases, there is always a trade-off between area, speed, and power.

Table 5.5 concludes that cache memory architecture having VMDSA with dual sleep technique and SRAMC with forced stack technique consumes the lowest power among all the SAs. Due to this, cache memory architecture having a VMDSA with dual sleep technique and SRAMC with forced stack technique has been simulated.

Table 5.6 describes that on applying PRT over different blocks of cache memory architecture such as SRAMC and VMDSA to optimize power consumption results depicted that SRAMC with forced stack technique and VMDSA with dual sleep technique in cache memory architecture consume the lowest power (8.078 µW) but increase in the number of transistors.

TABLE 5.4
Power Consumption of Cache Memory Architecture When $C = 6.09$ fF, and $R = 42.3$ KΩ on Applying Power Reduction Techniques Over SA

		Single-Bit SRAMC VMDSA Architecture		
S. No.	VMDSA Techniques	Power Consumption	Sensing Delay	Number of Transistors
1.	PRSTT	11.29 µW	13.51 ηs	32
2.	PRFST	11.29 µW	13.70 ηs	32
3.	PRSST	11.29 µW	13.50 ηs	33
4.	PRDST	11.03 µW	13.66 ηs	34

TABLE 5.5
Power Consumption of Cache Memory Architecture When C = 6.09 fF and R = 42.3 KΩ on Applying Power Reduction Techniques Over SRAMC

		Single-Bit SRAMC VMDSA Architecture		
S. No.	SRAMC Techniques	Power Consumption	Sensing Delay	Number of Transistors
1.	PRSTT	9.18 μW	13.12 ηs	32
2.	PRFST	9.10 μW	13.64 ηs	32
3.	PRSST	10.38 μW	13.36 ηs	33
4.	PRDST	10.13 μW	13.51 ηs	34

TABLE 5.6
Power Consumption of Cache Memory Architecture When C = 6.09 fF and R = 42.3 KΩ on Applying Power Reduction Techniques Over SRAMC and SA

		Single-Bit SRAMC VMDSA Architecture		
S. No.	SRAMC + VMDSA Techniques	Power Consumption	Sensing Delay	Number of Transistors
1.	PRSTT	9.27 μW	12.75 ηs	34
2.	PRFST	9.20 μW	13.14 ηs	34
3.	PRSST	8.46 μW	12.75 ηs	36
4.	PRDST	9.74 μW	13.34 ηs	38
5.	SRAMC with PRFST + VMDSA with PRDST	8.078 μW	12.14 ηs	36

5.11 CONCLUSION AND FUTURE SCOPE

In cache memory architecture, WDC, SRAMC and different types of sense amplifiers such as VMDSA, CMDSA, CTDSA, VLTSA, and CLTSA have been described in this chapter, as well as all architectures with different SAs implemented over cadence tool and power dissipation of all architectures have been calculated. In this chapter, cache memory architecture (CMA) with different types of SA such as VMDSA, CMDSA, CTDSA, VLTSA, and CLTSA has been implemented and compared on different values of resistance (R) with different parameters such as power consumption (PC), sensing delay (SD), and number of transistors (NT). Results depicted that cache memory architecture having VMDSA consumes the lowest power consumption (11.34 μW). Furthermore, power reduction techniques (PRT) such as sleep transistor technique (STT), forced stack technique (FST), sleepy stack technique (SST), and dual sleep technique (DST) are applied over different blocks of cache memory architecture to optimize power consumption, and the conclusion arises that SRAMC with forced stack technique and VMDSA with dual sleep technique in cache memory architecture consume the least power. Monte Carlo simulation and process corner simulation also have been done to check the robustness of the circuit. It is also observed that VMDSA has the lowest area 30.48 × 62.613 mm^2 as compared to other sense amplifiers. In the future scope, this work can be done in the form of an array. As conclusion arises that CMDSA with PRT consume low power but VMDSA consumes the lowest power among all SAs. So, in SRAMC array architecture researchers can use these amplifiers, and the power dissipation through the circuit can be further reduced by using other PRT and researcher also implemented SRAMC with forced stack technique in array in cache memory architecture.

REFERENCES

1. N. Eslami, B. Ebrahimi, E. Shakouri, and D. Najafi, "A Single-Ended Low Leakage and Low Voltage 10T SRAM Cell with High Yield," in Analog Integrated Circuits and Signal Processing, vol. 105, no. 2, pp. 263–274, 2020.
2. H. Bazzi, A. Harb, H. Aziza, M. Moreau, and A. Kassem, "RRAM-based Non-volatile SRAM Cell Architectures for Ultra-low-power Applications," in Analog Integrated Circuits and Signal Processing, vol. 106, no. 2, pp. 351–361, 2021.
3. S. Pal, S. Bose, and A. Islam, "Design of SRAM Cell for Low Power Portable Healthcare Applications," in Microsystem Technologies, vol. 14, pp. 1–2, 2020.
4. A. P. Shah, S. K. Vishvakarma, and M. Hübner, "Soft Error Hardened Asymmetric 10T SRAM Cell for Aerospace Applications," in Journal of Electronic Testing, vol. 36, pp. 255–269, 2020.
5. S. Singh, and V. Mishra, "Enhanced Static Noise Margin and Increased Stability SRAM Cell with Emerging Device Memristor at 45-nm Technology," in Radioelectronics and Communications Systems, vol. 61, pp. 200–206, 2018.
6. W. Wang, U. Guin, and A. Singh, "Aging-Resilient SRAM-based True Random Number Generator for Lightweight Devices," in Journal of Electronic Testing, vol. 36, pp. 301–311, 2020.
7. M. A. Khan, and K. A. Abuhasel, "Advanced Metameric Dimension Framework for Heterogeneous Industrial Internet of Things," in Computational Intelligence, vol. 37, no. 3, pp. 1367–1387, 2021.
8. Khaled Ali Abuhasel, and Mohammad Ayoub Khan, "A Secure Industrial Internet of Things (IIoT) Framework for Resource Management in Smart Manufacturing," in IEEE Access, vol. 8, pp. 117354–117364, 2020.
9. K. A. Abuhasel, and M. A. Khan, "A Secure Industrial Internet of Things (IIoT) Framework for Resource Management in Smart Manufacturing," in IEEE Access, vol. 24, no. 8, pp. 117354–117364, 2020.
10. H. Dounavi, Y. Sfikas, and Y. Tsiatouhas, "Periodic Aging Monitoring in SRAM Sense Amplifiers," 2018 IEEE 24th International Symposium on On-Line Testing And Robust System Design (IOLTS), Platja d'Aro, pp. 12–16, 2018.
11. A. Pathak, D. Sachan, H. Peta, and M. Goswami, "A Modified SRAM Based Low Power Memory Design," 2016 29th International Conference on VLSI Design and 2016 15th International Conference on Embedded Systems (VLSID), Kolkata, pp. 122–127, 2016.
12. Y. He, J. Zhang, X. Wu, X. Si, S. Zhen, and B. Zhang, "A Half-Select Disturb-Free 11T SRAM Cell with Built-In Write/Read-Assist Scheme for Ultralow-Voltage Operations," in IEEE Transactions on Very Large-Scale Integration (VLSI) Systems, vol. 27, no. 10, pp. 2344–2353, October 2019.

13. R. Fragasse et al., "Analysis of SRAM Enhancements Through Sense Amplifier Capacitive Offset Correction and Replica Self-Timing," in IEEE Transactions on Circuits and Systems I: Regular Papers, vol. 66, no. 6, pp. 2037–2050, June 2019.
14. S. Gupta, K. Gupta, B. H. Calhoun, and N. Pandey, "Low-Power Near-Threshold 10T SRAM Bit Cells with Enhanced Data-Independent Read Port Leakage for Array Augmentation in 32-nm CMOS," in IEEE Transactions on Circuits and Systems I: Regular Papers, vol. 66, no. 3, pp. 978–988, March 2019.
15. K. Sridhara, G. S. Biradar, and R. Yanamshetti, "Subthreshold Leakage Power Reduction in VLSI Circuits: A Survey," 2016 International Conference on Communication and Signal Processing (ICCSP), pp. 1120–1124, 2016, IEEE.
16. H. Jeong, T. W. Oh, S. C. Song, and S. O. Jung. "Sense-Amplifier-Based Flip-Flop with Transition Completion Detection for Low-Voltage Operation," in IEEE Transactions on Very Large Scale Integration (VLSI) Systems, vol. 26, no. 4, pp. 609–620, 2018.
17. Sunil Pandey, Shivendra Yadav, Kaushal Nigam, Dheeraj Sharma, and P. N. Kondekar, "Realization of Junctionless TFET-Based Power Efficient 6T SRAM Memory Cell for the Internet of Things Applications," Proceedings of First International Conference on Smart System, Innovations and Computing, Springer, Singapore, pp. 515–523, 2018.
18. K. Sridhara, G. S. Biradar, R. Yanamshetti. "Subthreshold Leakage Power Reduction in VLSI Circuits: A Survey," 2016 International Conference on Communication and Signal Processing (ICCSP), pp. 1120–1124, 2016, IEEE.
19. Reeya Agrawal, and V. K. Tomar, "Implementation and Analysis of Low Power Reduction Techniques in Sense Amplifier," Proceedings of the 2nd International Conference on Electronics, Communication, and Aerospace Technology (ICEC 2018), pp. 447–452.
20. J. Kathuria, M. A. Khan, A. Abraham, and A. Darwish, "Low Power Techniques for Embedded FPGA Processors," In: M. Khan, S. Saeed, A. Darwish, and A. Abraham (eds), Embedded and Real Time System Development: A Software Engineering Perspective. Studies in Computational Intelligence, vol. 520, Springer, Berlin, Heidelberg, 2014, https://doi.org/10.1007/978-3-642-40888-5_11.
21. Reeya Agrawal, and V. K. Tomar, "Analysis of Cache (SRAM) Memory for Core I™ 7 Processor," 9th International Conference on Computing, Communication and Networking Technologies (ICCCNT), 40225, 2018.
22. J. K. Mishra, H. Srivastava, P. K. Misra, and M. Goswami, "A 40 nm Low Power High Stable SRAM Cell Using Separate Read Port and Sleep Transistor Methodology," 2018 IEEE International Symposium on Smart Electronic Systems (iSES) (Formerly iNiS), Hyderabad, India, pp. 1–5, 2018.
23. Ziz Hao Liu, Tao Pan, Song Jia, and YuanWang, "Design of a Novel Ternary SRAM Sense Amplifier Using CNFET," 12th International Conference on ASIC (ASICON), pp. 207–210, 2017.
24. Gaurav Vashist, Hitesh Pahuja, and Balwinder Singh, "Design and Comparative Analysis of Low Power 64-Bit SRAM and Its Peripherals using Low-Power Reduction Technique," 5th International Conference on Wireless Networks and Embedded Systems (WECON), pp. 1–6, 2017.
25. John Keane, Jaydeep Kulkarni, Kyung-Hoae Koo, Satyanand Nalam, Zheng Guo, Eric Karl, and Kevin Zhang, "5.6 Mb/mm^2 1R1W 8T SRAM Arrays Operating down to 560 mV Utilizing Small-Signal Sensing with Charge-Shared Bitline and Asymmetric Sense Amplifier in 14 nm FinFET CMOS Technology," IEEE International Solid-State Circuits Conference, pp. 308–310, 2016.
26. YoungBae Kim, Qiang Tong, and Ken Choi, "Novel 8-T CNFET SRAM Cell Design for the Future Ultra-low Power Microelectronics," IEEE International SOC Design Conference (ISOCC), pp. 243–244, 2016.

27. Stefan Cosemans, and Francky Catthoor, "Comparative BTI Analysis for Various Sense Amplifier Designs," IEEE 19th International Symposium on Design and Diagnostics of Electronic Circuits & Systems (DDECS), pp. 1–6, 2016.
28. Bingyan Liu, Jiang Zeng Cai, Jia Yuan, and Yong Hei, "A Low Voltage SRAM Sense Amplifier with Offset Cancelling Using Digitized Multiple Body Biasing," IEEE Transactions on Circuits and Systems II, vol. 64, pp. 442–446, April 2016.
29. Mottaqiallah Taouil, and Said Hamdioui, "Integral Impact of BTI and Voltage Temperature Variation on SRAM Sense Amplifier," IEEE 33rd VLSI Test Symposium (VTS), pp. 1–6, 2015.
30. Byungkyu Jong, Taehui Na, and Jisu Kim, "Latch-Offset Cancellation Sense Amplifier for Deep Submicron STT-RAM," in IEEE Transaction on Circuits and System-I: Regular Paper, vol. 62, no. 7, pp. 1776–1784, July 2015.
31. Hanwool Jeong, Taewon Kim, and Kyomen Kang, "Switching PMOS Sense Amplifier for High-Density Low-Voltage Single-Ended SRAM," in IEEE Transaction on Circuits and Systems-I: Regular Paper, vol. 62, no. 6, pp. 1555–1563, June 2015.
32. C. A. Ajoy, A. Kumar, C. A. Anjo, and V. Raja, "Design and Analysis of Low Power Static RAM Using Cadence Tool in 180nm Technology," in International Journal of Computer Science and Technology, vol. 5, pp. 69–72, 2014.
33. J. P. Gajjar, A. S. Zala, and S. K. Aggarwal, "Design and Analysis of 32 Bit SRAM Architecture in 90 nm CMOS Technology," International Research Journal of Engineering and Technology, vol. 3, no. 04, pp. 2729–2733, 2016.
34. K. Vanama, R. Gunnuthula, and G. Prasad, "Design of Low Power Stable SRAM Cell," 2014 International Conference on Circuits, Power and Computing Technologies [ICCPCT-2014], pp. 1263–1267, 2014, IEEE.
35. Shikha Saun, and Hemant Kumar, "Design and Performance Analysis of 6T SRAM Cell on Different CMOS Technologies with Stability Characterization," in OP Conference Series: Materials Science and Engineering, vol. 561, 012093, 2019.
36. A. Bhaskar, "Design and Analysis of Low Power SRAM Cells," 2017 Innovations in Power and Advanced Computing Technologies (i-PACT), Vellore, pp. 1–5, 2017.
37. Yong-Peng Tao, and Wei-ping Hu, "Design of Sense Amplifier in the High-Speed SRAM," International Conference on Cyber-Enabled Distributed Computing and Knowledge Discovery, pp. 384–387, 2015.
38. Manoj Sinha, Steven Hsu, Atila Alvandpour, Wayne Burleson, Ram Krishnamurthy, and Shekhar Borhr. "High-Performance and Low-Voltage Sense-Amplifier Techniques for sub-90 nm SRAM," SOC Conference, 2003. Proceedings. IEEE International [Systems-on-Chip].
39. Ravi Dutt, and Abhijeet. "High-Speed Current Mode Sense Amplifier for SRAM Applications," in IOSR Journal of Engineering, vol. 2, pp. 1124–1127, 2012.
40. Yiqi Wang, Fazhao Zhao, Mengxin Liu, and Zhengsheng Han, "A New Full Current-Mode Sense Amplifier with Compensation Circuit," 2011 9th IEEE International Conference on ASIC, Xiamen, pp. 645–648, 2011.
41. L. Heller, D. Spampinato, and Ying Yao, "High-Sensitivity Charge-Transfer Sense Amplifier," Solid-State Circuits Conference. Digest of Technical Papers. IEEE International, 1975.
42. T. Na, S. Woo, J. Kim, H. Jeong, and S. Jung, "Comparative Study of Various Latch-Type Sense Amplifiers," in IEEE Transactions on Very Large-Scale Integration (VLSI) Systems, vol. 22, no. 2, pp. 425–429, February 2014.
43. Yiping Zhang, Ziou Wang, Canyan Zhu, Lijun Zhang, Aiming Ji, and Lingfeng Mao, "28 nm Latch Type Sense Amplifier Coupling Effect Analysis," 2016 International Symposium on Integrated Circuits (ISIC), Singapore, pp. 1–4, 2016.
44. D. Arora, A. K. Gundu, and M. S. Hashmi, "A High-Speed Low Voltage Latch Type Sense Amplifier for Non-volatile Memory," 2016 20th International Symposium on VLSI Design and Test (VDAT), Guwahati, pp. 1–5, 2016.

45. D. Schinkel, E. Mensink, E. Klumperink, E. van Tuijl, and B. Nauta, "A Double-Tail Latch-Type Voltage Sense Amplifier with 18ps Setup+Hold Time," 2007 IEEE International Solid-State Circuits Conference. Digest of Technical Papers, San Francisco, CA, pp. 314–605, 2007.
46. V. M. Tripathi, S. Mishra, J. Saikia, and A. Dandapat, "A Low-Voltage 13T Latch-Type Sense Amplifier with Regenerative Feedback for Ultra Speed Memory Access," 2017 30th International Conference on VLSI Design and 2017 16th International Conference on Embedded Systems (VLSID), Hyderabad, pp. 341–346, 2017.
47. A. Hemaprabha, and K. Vivek, "Comparative Analysis of Sense Amplifiers for Memories," 2015 International Conference on Innovations in Information, Embedded and Communication Systems (ICIIECS), Coimbatore, pp. 1–6, 2015.
48. M. Jefremow et al., "Time-Differential Sense Amplifier for Sub-80 mV Bit Line Voltage Embedded STT-MRAM in 40 nm CMOS," 2013 IEEE International Solid-State Circuits Conference Digest of Technical Papers, San Francisco, CA, pp. 216–217, 2013.
49. Y. Tao, and W. Hu, "Design of Sense Amplifier in the High-Speed SRAM," 2015 International Conference on Cyber-Enabled Distributed Computing and Knowledge Discovery, Xi'an, pp. 384–387, 2015.
50. S. Ahmad, B. Iqbal, N. Alam, and M. Hasan, "Low Leakage Fully Half-Select-Free Robust SRAM Cells with BTI Reliability Analysis," in IEEE Transactions on Device and Materials Reliability, vol. 18, no. 3, pp. 337–349, September 2018.
51. B. N. K. Reddy, K. Sarangam, T. Veeraiah, and R. Cheruku, "SRAM Cell with Better Read and Write Stability with Minimum Area," TENCON 2019 – 2019 IEEE Region 10 Conference (TENCON), Kochi, India, pp. 2164–2167, 2019.
52. A. Surkar, and V. Agarwal, "Delay and Power Analysis of Current and Voltage Sense Amplifiers for SRAM at 180 nm Technology," 2019 3rd International Conference on Electronics, Communication, and Aerospace Technology (ICECA), Coimbatore, India, pp. 1371–1376, 2019.

6 Investigation of Deep Learning Models for IoT Devices

Swagata Bhattacharya
Guru Nanak Institute of Technology
Kolkata, India

Debotosh Bhattacharjee
Jadavpur University
Kolkata, India

CONTENTS

6.1 Introduction 107
6.2 Overview of Deep Learning (DL) Models 109
    6.2.1 Cloud Offloading of DNN for IoT 112
    6.2.2 Architecture Trends for DNN 113
6.3 DNN Inference Engine Design for Low Resource Hardware Platform 113
    6.3.1 Available Hardware 114
    6.3.2 Software for DNN 114
    6.3.3 Hardware and Software Co-Design Procedure 115
6.4 Fitting DNN Model to IoT Devices 117
    6.4.1 Memory Modeling 117
    6.4.2 Data Flow Modeling 117
    6.4.3 DNN Architectures for IoT 119
        6.4.3.1 DNN for IoT 119
        6.4.3.2 Metrics and Performance Comparison of DNN Accelerators 123
6.5 Conclusion and Future Trends 124
    6.5.1 Approximate and Near-Threshold Computing 124
    6.5.2 Emerging Technologies 124
    6.5.3 Distributed Learning and Inference Distributed Computing 125
References 125

6.1 INTRODUCTION

Rapid development in IoT technology has led to an increasing number of connected sensory devices that collect or generate sensory data for vast fields of applications (Mohammadi et al. 2018; Sodhro et al. 2019). In the research done by CISCO, has been determined that there will be a 14.7 billion machine to machine (M2M) connections (which is also referred to as IoT) by 2023 (Cisco 2020). These IoT devices generate big data to be analyzed in real-time for smart applications (Mohammadi et al. 2018; Sodhro et al. 2017). Deep learning approaches (deep neural network or DNN) are known for their immense capability to handle big data (Najafabadi et al. 2015). DNN is very computation-intensive and requires significant time and energy. Deep learning is a subset of machine learning, which is again a thriving area in the field of artificial intelligence. "Machine learning is the field that gives computers the ability to learn without being explicitly programmed" as defined by Arthur Samuel. Machine learning designs are based on deep learning networks and are built on architecture with several layers of neurons grouped into layers by functionalities. A deep sequential structure is formed in DNN by organization of multiple layers.

In the emerging IoT applications, detecting a real-time response to user behavior and ambient conditions is of crucial importance. Extracting accurate inferences from raw sensor data is challenging within the noisy and complex environmental conditions where IoT systems are deployed. Deep learning has proved to be the most promising approach for overcoming the challenge and provides reliable inference (Deng and Yu 2014; Goodfellow, Bengio, and Courville 2015; Sheng et al. 2020). Emerging development in DNN techniques has become state-of-the-art for inference task increasingly in IoT applications like audio sensing and computer vision (speech (Hinton et al. 2012) and face recognition (Chen et al. 2014; Taigman et al. 2014)). Since deep learning algorithms impose depilating levels of system overhead in terms of memory, computation and energy, it is important to develop a systematic understanding of how DNN inference can be implemented and accelerated on resource-constrained platforms like IoT.

IoT can be seen as a communication between (1) man to man, (2) man to machine, and (3) machine to machine. To achieve these connections, the IoT is characterized (Ovidiu and Peter 2014) by:

1. *Interconnectivity:* This refers to the property of IoT that everything can be interconnected globally with communication infrastructure.
2. *Things-related services:* It provides privacy protection between physical and virtual things.
3. *Heterogeneous:* IoT devices show heterogeneity as they can interact with different kinds of devices and service platforms.
4. *Dynamic changes:* The state and number of devices changes dynamically with reference to sleep and wake mode or connected-disconnected.
5. *Security and safety:* These vital issues are addressed in IoT devices and it includes personal data and physical safety of man and devices.
6. *Connectivity:* This includes the ability of the devices to get access to network and common ability to consume and produce data.
7. *Enormous scale:* It refers to management and interpretation of generated data for various applications.

The IoT is therefore a marriage between computing and communication and creates the opportunity for advanced applications, such as smart home, smart building, and smart cities. For such scenario an embedded design should produce a result in

TABLE 6.1
Some DNN Framework and Their Hardware Support

Software Frameworks	References	Supported API	Supported System	Supported Hardware
TensorFlow	Abadi (2016)	C++, Python	Linux Windows OS	CPU, GPU, Raspberry PI
Pytorch	Vohra and Fasciani (2019)	Cuda, C/C++	Linux Windows OS	GPU
Deeplearning 4J	Alwzwazy (2016)	Java, CUDA	Linux Windows OS	GPU
Caffe	Jia et al. (2014)	C++, Python, MATLAB	Linux Windows, macOS, Rasberian OS	CPU, GPU
MxNET	Chen et al. (2015)	C++, Python, JAVA, JULIA, PERL, SCALA	Linux Windows OS, Rasbian OS	CPU, GPU, Raspberry PI, NVIDIA JETSON
Keras	Jakhar and Hooda (2018)	Python	Linux Windows OS	CPU, GPU, TPU
DarkNet	Redmon (2013)	C++, Python	Not specified	CPU, GPU
SCALE-SIM: Systolic array DNN accelerator simulator	Samajdar et al. (2018)	C code and HLS	Not specified	Google TPU ASIC

seconds-not in hours; on the other hand, a data center could not have a power plant of its own, so energy consumption should be optimized even if the accelerator is designed for delivering high performance. Naturally, these constraints pose several challenges when it comes to making design choices. Deploying DNN in resource-limited IoT devices is, therefore, a challenging area of research. The main challenge in deploying a deep learning model within a resource-constrained environment such as IoT devices or smartphones is that it becomes critical to find a balance between model accuracy and computational cost to ensure the model will function well within resource-limited environments. A DNN framework is shown in Table 6.1.

6.2 OVERVIEW OF DEEP LEARNING (DL) MODELS

DL is a branch of machine learning which is based on many artificial neural networks (Mohammadi et al. 2018). The purpose of ANN is to act like a human brain, so DL also mimics the human brain (Mohammadi et al. 2018). Its architectures consist of multiple processing layers: an input layer, hidden layers, and an output layer. The basic building block of DNN is a neuron. A neuron receives several inputs and performs a weighted summation of those inputs. The summation result is put to an activation function to produce output. Each neuron inputs are associated with weights and bias that is required to be optimized during training.

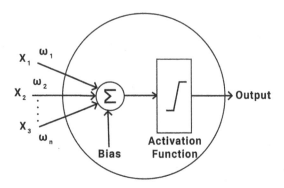

FIGURE 6.1 Basic structure of a neuron. (Based on Mohammadi et al. 2018.)

Figure 6.1 shows the structure of a neuron. During the training period, the input layer assigns (usually randomly) weights to the data and it is passed to next layer. Weights are also assigned by the subsequent layer to the input and produce output which is fed as input to the following layers. The model prediction is produced at the last layer. The prediction's correctness is determined by the loss function by computing error rate between predicted and actual value. Backpropagation algorithm is used to propagate error rate across the input layer. In each cycle weights are balanced. The training cycle is repeated until below threshold value is attained by error rate.

Figure 6.2 shows the high-level mechanism of training for DNN.
The popular deep learning models are:

1. Classic neural network (multilayer perceptron):
   The classic neural network consists of more than two hidden layer or have multiple layers of perceptron. This network is mostly used for classification and regression where a dataset is given as input. It's also used where a high level of flexibility is required.

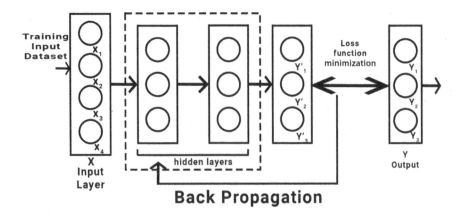

FIGURE 6.2 The overall mechanism of the DL model.

2. Convolutional neural network (CNN):
   CNN is the most stunning architecture of DL technique. It can deal with greater complexity around pre-processing and data computation. It has six major building blocks:
   1. *Input layer:* As all ANNs, the input layer holds the image pixels.
   2. *Convolution layer:* In this step, feature maps are created for input dataset. When the data hits a convolutional layer, the layer convolves each filter across the input's spatial dimensionality to produce a 2D activation map.
   3. *Max-pooling:* Pooling layers aim to reduce the representation's dimensionality gradually, thus further reducing the number of parameters and the model's computational complexity. It down samples the outputs of the previous convolutional layer. It helps the CNN to detect a modified image.
   4. *Fully connected layer:* Its role is the same as that of fully connected layers in other ANNs and the class scores generated here is used for classification.
   5. *SoftMax/logistic layer:* Logistic is used for binary classification, and SoftMax is for multi-classification.
   6. *Output layer:* The output layer contains the label, which is in the form of a one-hot encoded.

   In CNN (Figure 6.3), the time complexity at the training stage is improved compared to fully connected layers because each neuron is connected to a small subset of input. This decreases the total parameters in the network. This property is called local connectivity. CNN finds exciting applications in IoT devices, such as drones, smartphones, and smart transports equipped with cameras. The CNN architecture and its variations have been investigated for various application scenarios that involve these devices. CNN is currently used for developing smart homes (Njima et al. 2019) and cities facility and precision agriculture. Although CNN was built for image data, it also shows remarkable result in non-image data.

3. Recurrent neural network (RNN):
   Though the CNN is considered as a most powerful tool in the field of computer vision (Heet et al. 2015) but both CNN and DNN (Yu 2019) suffer from the problem that they cannot deal with temporal information of input data. So, the research on sequential data such as text, audio, and video uses RNN. RNN consisting of standard recurrent cells (e.g., sigma cells) have had incredible success on some problems. But RNN has the inherent problem of vanishing gradient. This means that long-term information has to sequentially travel through all cells before getting to the present processing cell. This means it can be easily corrupted by being multiplied many times by small numbers <0. This major limitation is addressed by LSTM.

FIGURE 6.3 The flow of CNN.

4. Long short-term memory (LSTM):
   In order to handle the "long-term dependencies," LSTM was proposed by Hochreiter and Schmidhuber (1997). They introduced an input gate into the cell. The attention model is also an advancement in RNN.
5. Auto encoder (AE):
6. AEs work by automatically encoding data of inputs, perform on activation function, and finally decoding data for output. Denoising AE has been extended from auto encoder as asymmetrical neural network for learning features from noisy datasets.

There are other models like Boltzmann machines and deep belief network which are beyond the scope of this chapter.

6.2.1 Cloud Offloading of DNN for IoT

Offloading to the cloud was chosen to deal with power and performance constraints in some experiments in literature. A CNN-based object-inference task on an Nvidia Jetson TX1 device was used to compare its performance and power consumption to the case of offloading these services to the cloud. Images were sent to the cloud and then waited for the responses. Tang et al. (2017) showed that, for object recognition, executing locally consumes 7 W compared to 2 W when offloading to the cloud. This shows that offloading is indeed an effective way to reduce power consumption. However, offloading also leads to a minimum latency of 2s, and latency can be as high as 5s—which doesn't meet the real-time requirement of 500 ms. This led to the conclusion that with current Internet speeds, offloading to the cloud isn't yet a practical and acceptable solution for real-time deep-learning tasks (Tang et al. 2017). Three main problems to execute DNN through the cloud are (Iandola and Keutzer 2017; Li et al. 2018):

1. *Latency:* A significant drawback of real-time processing of data on the cloud is latency (Sze et al. 2017). IoT devices are long-distance and have low-bandwidth communications with the cloud, which leads to unacceptable latency. For example, a fast DNN named YOLO (Ndikumana et al. 2020) was initially designed to achieve a benchmark speed of 45 frames per second (FPS) on a powerful Titan X GPU. But the speed lowered (15 FPS) on cloud-chaser (Redmon et al. 2016).
2. *Security/privacy:* A risk of unauthorized access to sensitive information exists during communications between IoT devices and the cloud. A person's video may be intercepted in a video surveillance system while communicating through the cloud, posing a significant hindrance to security (Luo et al. 2019). So, the security solutions developed (Al-Garadi et al. 2020; Li et al. 2018; Stergiou et al. 2018) are not as per the requirement.
3. *Dependence on wireless connection:* Running DNNs on the cloud requires dependence on wireless connections. In real-time systems where response time is crucial, like automated cars, it may pose life risk if data on the cloud is not processed in a specified time in case of poor wireless

Investigation of Deep Learning Models for IoT Devices 113

connectivity. It is preferred to deploy real-time data processing on the IoT device. Mohammadi et al. (2018) and Wu et al. (2017) developed a fully convolutional DNN, named Squeezedet, which could be implemented on embedded devices to detect objects for real-time autonomous driving. The speed of 57.2 FPS was achieved, benefitted from a small size of 7.9 MB.

All the reasons mentioned above have driven the research community to find efficient ways to deploy DNN in IoT devices.

6.2.2 Architecture Trends for DNN

The architecture evolution of deep learning devices was coupled with the machine learning community's algorithmic changes. In the early days, GPUs were used to accelerate the central computational part of CNN. It was possible because the early CNN structure's computational requirement matched with the GPGPUs SIMD architecture and floating-point utilization. As CNN became less regular, the research point was shifted to a new requirement. Soon the Nvidia Tesla V100 came with Tensor Core Technology to accelerate AI work. New ASIC such as Tensor Processing Unit (TPU) appeared from Google and provided fixed-point calculation. Lake Crest Chip from Intel supports floating-point calculation, another ASIC for accelerating AI. In the embedded systems field, low power ASICs have arrived like the Myriad 2 chip from Movidius and projects from academia such as Eyeriss (Chen, Emer, and Sze 2016) and Cnvlutin (Albericio et al. 2016). Field-programmable gate arrays have also gained much attention due to their less turnaround time, re-programmability and low power and portability, and competing CPU and GPU counterparts under a power budget. Many software tools have been developed to help researchers map CNNs on the above devices without extensive knowledge of architecture and develop high-performing systems. The academia has developed several tools such as fpgaConvNet (Venieris and Bouganis 2016), FP-DNN (Y. Guan et al. 2017), and FINN (Umuroglu et al. 2017). The cuDNN from NVIDIA for Cuda enabled GPU, Xilinx's targeted FPGGAs have scope for DL processing and accelerating. The application-specific hardware provides the best efficiency in terms of the power budget.

6.3 DNN INFERENCE ENGINE DESIGN FOR LOW RESOURCE HARDWARE PLATFORM

DNN has two phases: (i) training and (ii) inference. There are significant computation steps in the training phase because of multiple iterations in forward and backward calculations of DNN. So, the training phase must be implemented at a robust resource center rather than resource-constrained devices. Hence, the inference stage is implemented by researchers on IoT devices. There is a significant memory space and memory access requirement for the inference stage, which is at the cost of significant energy consumption. Therefore, it is challenging to implement DNN's inference stage in low resource hardware devices in real-time. Researchers have been developing energy efficient hardware accelerator architectures, to fulfill the market need. Besides, model optimizations have been researched considering the trade-off among model size (arithmetic precisions, number of parameters, and operations), recognition accuracy, speed, and energy cost.

6.3.1 Available Hardware

FPGA (Zhang and Kouzani 2020) has emerged as the most efficient hardware platform in this era for accelerating DNN due to its low cost, less turnaround time, high speed, energy efficiency, parallelism, and programmability in architecture (Guo et al. 2017; Van Gerven and Bohte 2017; Venieris, Kouris, and Bouganis 2018). FPGA hardware platform contains an FPGA chip, on-board off-chip memory, interfaces, and peripherals. An FPGA can be connected to the host server through PCI (peripheral component interconnect) express.

Some platforms have integrated processing units like the Xilinx Zync 706 and Pynq-Z1 integrate dual ARM Cortex-A9 CPUs, and Zynq Ultra Scale MPSoC ZCU 102 contains Mali-400 GPU and ARM V8-based quad-core multiprocessing CPU. The basic architecture of FPGA has an array of configurable logic blocks (CLBs), input/output (I/O) blocks, transceivers, and other interconnection resources to support on-chip (interblocks) communications, and other operations (Bajaj 2016; XILINX 2021).

A block dedicated for DNN computation on FPGA and ASIC is referred to as processing element (PE) (Chen, Emer, and Sze 2016). FPGAs' properties are directly related to IoT devices' energy consumption and throughput. The FPGAs are grouped into four categories (Zhang and Kouzani 2020): low-resource FPGAs, medium-resource FPGAs, high-resource FPGAs, and ultra-high-resource FPGAs (Luo et al. 2019).

The metrics that measure the FPGA's capability to run a DNN algorithm at the inference stage are chip-RAM memory capacity, aggregate bandwidth of transceivers, I/O count. Since memory requirement for computation-intensive DNNs is a primary concern, other parameters depend on on-chip memory for memory access and data transfer.

Apart from FPGA platforms, Xilinx has developed Omnitek's deep learning processing unit (DPU) (Vipin and Fahmy 2018). It supports 800 MHz DSP performance with 86% efficiency. Intel and Google have developed the vision processing unit (VPU) and tensor processing unit (TPU) for accelerating the on-device DNN inference.

Intel's Nervana Neural Network Processor (NNP) proved to have better efficiency than CPU processors regarding acceleration of both training and inference of DNNs. In this regard, FPGA is the choice for its low-cost, lower turnaround time, and field programmable features. In this chapter, we focus on real-time low-energy DNN inference using FPGA-based platforms or dedicated chips.

6.3.2 Software for DNN

In Table 6.2, we summarize some current popular software frameworks written in C, C++, Python, and Java. TensorFlow is the most popular framework and it is available in desktop as well as mobile devices that have Android or Raspbian. It has multiple GPU support. It is well known for Google Translate and has natural language processing capability. Keras provides excellent rapid prototyping and parallel programs that

TABLE 6.2
Some Popular DNN Models and Their Hardware Targets

DNN Model	Accelerator	Reference	Target	Year	Hardware Support
Convolutional Neural Network (CNN)	Approximate Accelerator using Caffe	Castro-Godínez et al. (2020)	FPGA (SoC-based IoT)	2020	ARM Cortex-A9, 32-bit, 0.8 GHz, 1 GB RAM
	Heterogeneous processor with accelerator	Liu et al. (2020)	FPGA	2020	Xilinx XC7VX690T FPGA with RISC-V
	Compact CNN accelerator	Ge et al. (2019)	FPGA	2019	ARM, Cortex-M3 (CM3)
Recurrent Neural Network (RNN)	RNN Fast	Samavatian et al. (2020)	Mobile GPU	2020	Zynq 7020 FPGA from Xilinx
	BRAINWAVE	Fowers et al. (2018)	FPGA	2018	Intel Stratix 10 280 FPGA
Long Short-Term Memory (LSTM)	ELSA	Azari and Vrudhula (2020)	ASIC	2020	65 nm CMOS technology
	POLAR	Bank-Tavakoli et al. (2019)	FPGA	2019	Xilinx XC7Z045 device
	FPGA accelerator	Guan, Yuan et al. (2017)	FPGA	2017	Xilinx VC707

can be implemented in multiple GPUs. It is capable of constructing DNN and executing tensor computation. It has been enhanced with TensorFlow serving tool that has the capacity to deploy any DNN algorithm while maintaining the same architecture and APIs. Deeplearning4J is adopted as commercial, industry oriented, and distributed DL platform. It supports CNN, RNN, and LSTM. It finds matchless results for image recognition, fraud detection, and text mining. In comparison, MNet supports Raspberry Pi and NVIDIA Jetson devices. Moreover, OpenCL Caffe has been developed to support AMD and Intel devices specifically. The DNN models generated using various tools can be translated into ONNX format and then transferred from one framework to another using ONNX. To efficiently execute DNN inference on a hardware platform using these software frameworks, Intel and Xilinx have developed toolkits to support their devices (e.g., FPGAs and VPU), i.e., Xilinx in-depth neural network development toolkit (DNNDK) and Intel Openvino toolkit. DNNDK is initially designed for executing DNNs on a host CPU with Xilinx DPU and FPGA platforms. DNNDK supports both high-level frameworks (i.e., TensorFlow and Caffe) and low-level framework of Darknet on Linux or Windows-based systems.

6.3.3 Hardware and Software Co-Design Procedure

The brute-force computing model of DNN often requires vast hardware resources, introducing severe concerns on its scalability running on traditional von Neumann architecture. The well-known memory wall and latency brought by the long-range

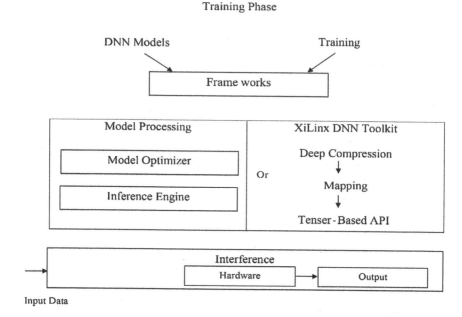

FIGURE 6.4 Hardware-software co-design framework. (Modified from Zhang and Kouzani 2020.)

connectivity and communication of DNN severely constrain the computation efficiency of DNN. The acceleration techniques of DNN, either software or hardware, often suffer from low hardware execution efficiency of the simplified model (software) or inevitable accuracy degradation and limited supportable algorithms (hardware), respectively. To preserve the inference accuracy and make the hardware implementation more efficient, an immediate investigation of the hardware/software co-design methodologies for DNNs is needed.

A generic hardware/software co-design process is shown in Figure 6.4. Using a high-level framework (e.g., TensorFlow), a DNN model can be built and trained using a training set.

During the training process, the DNN model can be optimized from scratch using its sparse structure. On the other hand, the model optimizer can be used to design, decompose, prune, and quantize, e.g., DeepCompression Tool (Intel 2020). A compiler (e.g., Xilinx DNN compiler) then converts the compressed DNN model into binary code using suitable mapping and implementing it on FPGA.

Meanwhile, the memory hierarchy optimization is required to support the dataflow mapping strategy. Finally, the DNN inference can be executed on an IoT device through APIs to process the real-time data. The parameters like recognition accuracy, number of parameters and MAC operations, speed, and power consumption have to be considered. Next, during the development (including training and inference) of a DNN, the following parameters need to be considered: recognition accuracy, number of parameters and MAC operations, energy consumption, and speed.

Investigation of Deep Learning Models for IoT Devices

The aim is to fit the DNN into a low-power and limited resource hardware platform for a real-time application. This described process is called hardware-software co-design. It is essential that the co-design approach can utilize the popular tools that can be compatible with tools. The co-design approach is depicted in Figure 6.4.

6.4 FITTING DNN MODEL TO IoT DEVICES

It is a challenging research area to fit DNN into energy sensitive IoT devices that provide limited on-chip memory and computation capability. A DNN model can be optimized in two ways: (i) optimizing from scratch: where an optimizing method is developed in the training phase to explore sparsity of model; (ii) using model optimizer: It is designed to decompose, prune, and quantize already trained DNN. A DNN compiler with a mapping strategy then converts the DNN model to binary code and then into FPGA. The two main factors that are most important for energy consumption and acceleration performance are (i) memory modeling and (ii) data-flow modeling.

6.4.1 MEMORY MODELING

Memory comes out as the major bottleneck in implementing DNN models in low-resource devices like IoTs. FPGA is usually the most acceptable hardware choice but its off-chip memory has bandwidth limitation. Initially when the neural networks didn't require large memory for activation, then FPGA architecture on-chip memories were well sufficient for such structures. An extensive set of data is usually stored in a DRAM (Chen, Emer, and Sze 2016; Guan, Liang et al. 2017) to reduce the expenditure, and that a small set of data is stored in the register file (RFs) or an on-chip buffer. In Chen, Emer, and Sze (2016), the off-chip memory architecture using DRAM is described. In FPGAs, the on-chip buffer can be built by multiple distributed block RAMs to store a large number of Ifmaps and filter channels. An extensive set of data is usually stored in a DRAM (Chen, Emer, and Sze 2016; Guan, Liang et al. 2017) to reduce the expenditure, and that a small set of data is stored in the register file (RFs) or an on-chip buffer. Also, there are two types of data, i.e., Ifmaps, filters, and Ofmaps. Ofmaps are also known as partial sums (Psums). But as the complexity of DNN grows like in GoogleNet, ResNet, and DenseNet they exhibit diversity in computation and communication resulting in bandwidth requirement higher than the device. It has become critical to decide which data should be used by on-chip memory to maximize performance and memory usage. So on-chip memory allocation strategy is very important to have proper memory management. The techniques that are adopted to solve the memory bottleneck are (i) spatial data reuse, (ii) hierarchical memory, and using (iii) exploiting data locality.

6.4.2 DATA FLOW MODELING

In Chen, Emer, and Sze (2017), an analogy was explained between DNN acceleration and general-purpose processing. It showed that in normal computer, while the compiler generates machine-readable binary codes from the program for execution; in DNN accelerator, the mapper translates the DNN shape and size into a hardware-compatible mapping for execution. The compiler usually optimizes for performance; the mapper significantly optimizes for energy efficiency (Chen, Emer, and Sze 2017).

The data reuse technique for memory management uses three major dataflow topologies.

This part of the chapter describes these dataflow models required for DNN execution (Chen, Emer, and Sze 2017) as depicted in Figure 6.5.

Weight-stationary dataflow: In this scheme, the filter weights are fetched once in each processing element (PEs) and multiplied with many input values. Finally, the output psums are accumulated across Pes. Thus, dataflow topology shows parallelization of input data while the weights remain stationary. Thus, though the weight bandwidth is improved but at the cost of input memory bandwidth.

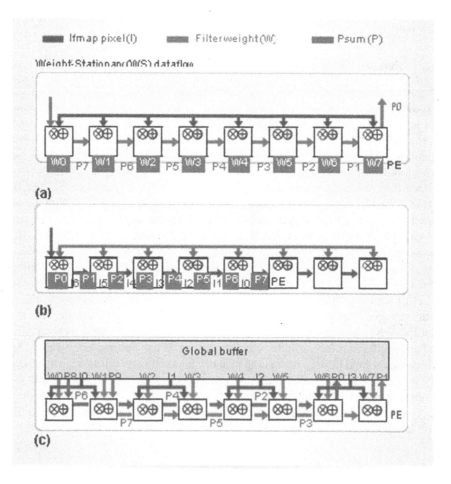

FIGURE 6.5 Dataflow taxonomy (a) weight stationary, (b) output stationary, and (c) no local reuse. (Reproduced from Chen, Emer, and Sze (2017) with permission.)

Investigation of Deep Learning Models for IoT Devices

Output-stationary (OS) dataflow: In this scheme, the filter weights and inputs are loaded in every cycle in the PEs but accumulate the intermediate results or psums in the MAC units. The mapping rule is that all MACs that generate psums for the same Ofmap pixel must be mapped on the same PE serially. This enables to maximize the output data bandwidth and output data reuse (see Figure 6.5(b)).

No-local-reuse (NLR) dataflow: This dataflow keeps no data stationary locally. This is to minimize DRAM access energy consumption by storing more data on-chip (Chen et al. 2014; Zhang et al. 2015). The corresponding mapping rule is that all parallel MACs must come from a unique pair of filters and channels at each processing cycle (see Figure 6.5(c)).

The three dataflows show distinct data movement patterns, which imply different trade-offs.

Energy efficient dataflow: Row-stationary (RS), which aims to optimize data movement for all data types in all levels of the storage hierarchy of spatial architecture.

The RS dataflow divides the MACs into *mapping primitives,* each of which comprises a subset of MACs that run on the same PE in a fixed order. Specifically, each mapping primitive performs a 1D row convolution, so we call it a *row primitive,* and intrinsically optimizes data reuse per MAC for all data types combined.

6.4.3 DNN Architectures for IoT

This section gives an overview of the latest DNN hardware architecture methods implemented for IoT devices. The chapter limits exploring only to CNN and LSTM architectures. Some popular DNN models and their hardware support are summarized in Table 6.2.

6.4.3.1 DNN for IoT

Two approaches have become the trend for deploying CNN on IoT platforms: (i) scaling down a large model and (ii) designing a small model especially targeting the resource-constrained platform (Lawrence and Zhang 2019). Some popular DNN models and their hardware target and support are summarized in Table 6.2. For these accelerator designs, Vivado Tool and Cadence Tool were intensively used. In this section, we also explore some state-of-the-art DNN implementations in hardware accelerators as explained below.

a. **IoTNet:** IoTNet (Lawrence and Zhang 2019) is a small efficient model and shows immensely state-of-the-art performance as it is explicitly designed for a resource-constrained environment. Instead of performing depth-wise convolution, this work simplifies the process as IoTNet factorizes standard 3×3 convolutions into pairs of 1×3 and 3×1 standard convolution. This reduces number of parameters by 33%. Thus, it makes a tradeoff between the cost of computation and accuracy. IoTNet has scaled-down large architectures on datasets that best match the complexity of problems faced in resource-constrained environments. The work compares model accuracy and the number of floating-point operations (FLOPs) to measure the

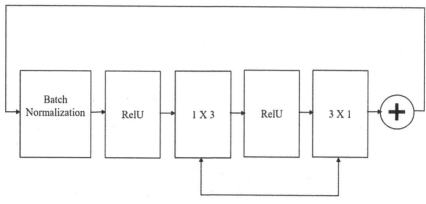

1 X 3 and 3 X 1 Standard Convolution

FIGURE 6.6 IoT net network flow.

efficiency. It reports accuracy improvement over MobileNetV2 on CIFAR-10 of 13.43% with 39% fewer FLOPs, over ShuffleNet on Street View House Numbers (SVHN) of 6.49% with 31.8% fewer FLOPs.

It has an initial 3 × 3 convolutions, followed by at least one group of blocks. Figure 6.6 shows the block diagram. A block consists of batch normalization (Zagoruyko and Komodakis 2016) followed by a pair of 1×3 and 3×1 standard convolutions and contains a skip connection. F is a residual function, and Wl is the parameter matrix of the block. Each convolution is preceded by a ReLU (Glorot, Bordes, and Bengio 2011).

b. **Low power CNN architecture for IOT device:** In Elgawi and Mutawa (2020), a DeepIoT low-power CNN accelerator is designed, and the figure shows a low power framework implemented in FPGA.

It includes two techniques: (1) a model compression is attained using a hybrid quantization scheme so that different bit-widths precision can be applied across CNN layers (Wang et al. 2018) and (2) an activation function based on the recast of softmax layer with a linear support vector machine (SVM) (Tang 2013).

This work tries to have a trade-off between power consumption and network model size. Its significant contributions are: (1) to maximize throughput on resource-constrained IoT devices, (2) to present a dynamic bit-width precision data quantization scheme to maximize the energy efficiency of the CNN inference, and (3) to develop an automatic design flow that automatically generates a synthesizable accelerator from a high-level specification. An overview of DeepIoT architecture is shown in Figure 6.7.

Hybrid Bit-width Quantization Architecture: With the hybrid quantization scheme, binary weights are used in FC layers for significant gains on the Ed while using ternary weights for the CONV layers to reduce complexity on the Ec. These flexible quantization schemes help build resource-customized and energy efficient CNN accelerators, not affecting inference performance.

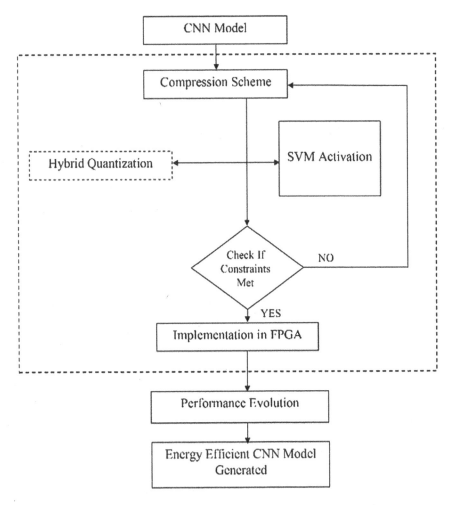

FIGURE 6.7 Overview of DeepIoT architecture. (Inspired from Elgawi and Mutawa (2020.)

Quantization allows baseline CNN models, such as VGG-16, to efficiently perform inference on the fixed-point pipeline, thus getting increasingly popular in mobile devices. A hybrid quantization scheme is used to quantize the network model parameters into different bit-widths precision. Bit-width precision quantization makes the precision of computation adaptive to various requirements while improving CNN's energy efficiency. By using flexible dynamic bit-width precision quantization, the energy efficiency of CNN inference is improved, while keeping computational precision adaptive to varying requirements.

c. **RAPIDNN:** In RAPIDNN (Imani et al. 2018), all functionalities are mapped inside the memory block. It is the first neural network accelerator of this kind. RAPIDNN ensures scalability as it uses direct digital-based computation without the requirement of analog-to-digital conversion.

RAPIDNN is implemented using commonly used single-level memristor devices; thus, need for multi-level memristor is removed.

This accelerator's salient features are:

- Software support for RAPIDNN and with novel algorithms which reinterpret DNN models to enable in-memory processing with minimal accuracy loss of DNN inference with DNN inference shows minimum accuracy loss.
- Users can configure RAPIDNN for different DNN applications because adjustable DNN reinterpretation mechanisms are provided.
- RAPIDNN can provide the same prediction quality level as demonstrated in six applications using small-sized memory blocks while ensuring less than 0.5% of quality loss.

Figure 6.8 illustrates an overview of the RAPIDNN framework.

- It consists of two interconnected blocks: A software module, a DNN composer, and a hardware module, an accelerator.
- The DNN composer converts each neural network operation to a table stored in the accelerator memory blocks.
- This table processes all neural network computations inside the memory. The work employs a step-wise function approximation technique to form input-output tables that replaces CMOS logic units of processors.
- Input-output is analyzed offline.
- DNN composer starts with a DNN model, analyzes each neuron's weights and inputs, and generates a more suitable DNN appropriate for the PIM accelerator.
- The final model is generated after repeated iterations and stored in an accelerator for performing online inference.

d. **SCALE-SIM:** This work comes with two major features. Firstly, it describes a cycle-accurate simulator called SCALE-SIM for DNN inference on systolic arrays (systolic array-based DNN simulator). This is used to model

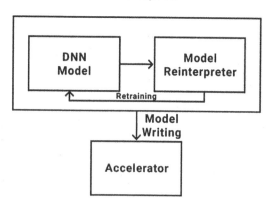

FIGURE 6.8 Overview of RAPIDNN architecture. (Based on Imani et al. (2018).)

both scale-up and scale-out systems, modeling on-chip memory access, runtime, and DRAM bandwidth requirements for a given workload. Secondly, it is an analytical model to estimate the optimal scale-up vs. scale-out ratio given hardware constraints (e.g., TOPS and DRAM bandwidth) for a given workload.

e. **FPGA/DNN co-design:** An efficient design methodology for IoT: This is a fully automated co-design flow and first work in this co-design methodology. A bottom-up approach is maintained following a hardware-oriented DNN model. At the same time, DNN-driven FPGA accelerator may be designed using a top-down approach.

- For DNN model design introduced to guide the DNN generation with well ahead predictable performance that reduces the co-design search space. Based on such a template, an automatic DNN model search engine, Auto-DNN, is proposed.
- For FPGA accelerator design, fine-grained tile-based pipeline architecture is introduced in this work which supports arbitrary DNNs generated by Auto-DNN using a library of highly optimized HLS IPs. Based on such architecture, an automatic HLS generator, Auto-HLS, is proposed to directly generate synthesizable C code of the DNN models to conduct latency/resources estimation and FPGA accelerator generation.
- The co-design approach applied on object detection task implemented in PYNQ-Z1 embedded FPGA. DNN models are searched and mapped to the board with the state-of-the-art performance regarding the accuracy, speed, and power efficiency.

6.4.3.2 Metrics and Performance Comparison of DNN Accelerators

Researchers follow some standard metrics to define the hardware's performance that refers mainly to the area, power consumption, and operations per second.

- *Area:* The area, generally expressed in mm^2 or micrometer square, represents the portion of silicon required to contain all the necessary logic. It strictly depends on the technological node used during the hardware synthesis process and the on-chip memory size.
- *Power:* The power consumption comes from the device's power requirement and the application for which it was designed. Both on-chip and off-chip memories' power requirement are also included in power consumption as they are the primary source.
- Throughput defines how often an accelerator can accomplish a complete convolution or a complete inference. Usually, it is expressed as billions of operations per second (Gop/s) or billions of Macs per second (GMAC/s). Since accelerators performance is application-dependent, so very often, comparisons between them are complicated and should be evaluated based on datasets or standard models.

Table 6.3 makes a comparison between the explained DNN accelerators above in IoT devices.

TABLE 6.3
Comparison between DNN Accelerators in IoT Devices

Name	Platform	Reference	Technology (nm)	Area	Power	Gop/s
Neural Network Accelerator	FPGA Altera Cyclone V	Hong and Park (2017)	28 nm	–	413.99 mW	11.95
Low-power deep learning architecture	FPGA Xilinx Virtex 7	Elgawi and Mutawa (2020)	28 nm		155.3W	
RAPIDNN	HSPICE Synopsys Design Compiler	Imani et al. (2018)	–	124.1mm^2	155.3W	839.1 GOP/s/W
FPGA/DNN Co-design (Tile-Arch Accelerator)	PYNQ-Z1 FPGA	Hao et al. (2019)		–	2.2W	–
SMIV	SoC	Véstias (2020)	16 nm	25 mm^2		

6.5 CONCLUSION AND FUTURE TRENDS

This chapter discussed the challenges and opportunities for implementing DNN in resource-constrained hardware like IoT. The future directions in this regard are discussed in the subsections that follow.

6.5.1 APPROXIMATE AND NEAR-THRESHOLD COMPUTING

Approximate and near-threshold computing are emerging techniques for designing ultra-low-power computing devices. Quality is sacrificed to obtain improvement in area, power, performance, and/or energy. These techniques have shown remarkable results in designing resource-efficient systems. But they are yet to take hold of the resource-efficient DL systems for near-sensor computing in IoTs.

6.5.2 EMERGING TECHNOLOGIES

Nano-wires, Memristors, Spintonic (University of Minnesota 2013) devices, etc., are anticipated to be the beginning of a new computing era. These technologies are expected to offer highly resource-efficient solutions that can help improve edge devices' computational capability under a defined energy/power or performance constraint. Nano-wires-based 3D-stacked architectures are examples of such futuristic devices, where a sea of accelerators is connected with a high-speed dense memory through high-bandwidth nano-wires.

6.5.3 DISTRIBUTED LEARNING AND INFERENCE DISTRIBUTED COMPUTING

It is also famous for simulating neural networks in IoTs-based systems(Cuervo et al. 2010). The idea behind distributed computing is to ensure fine-grained energy latency aware distribution from edge to the cloud in IoTs. This reduces energy consumption in near sensor devices and achieves the system's overall performance improvement.

This chapter concludes that continuous development in framework and hardware for DNN execution in IoT devices shows a tremendous need globally for such research. This chapter summarizes some very new approaches to deploying DNN in IoTs and comparing their performance based on standard metrics. New techniques are needed to be explored to make the DNN research for IoTs on track with emerging trends in technologies when Moor's law seems to have an end.

(Abadi et al. 2016; Albericio et al. 2016; Al-Garadi et al. 2020; Alwzwazy et al. 2016; Aly et al. 2015; Azari and Vrudhula 2020; Bajaj 2016; Bank-Tavakoli et al. 2019; Bottou 2010; Castro-Godínez et al. 2020; Chauvin and Rumelhart 1995; Chen et al. 2014; Chen et al. 2015; Chen, Emer, and Sze 2016; Chen, Emer, and Sze 2017; Cisco 2020; Cuervo et al. 2010; Deng and Yu 2014; Elgawi and Mutawa 2020; Fowers et al. 2018; Ge et al. 2019; Glorot, Bordes, and Bengio 2011; Goodfellow, Bengio, and Courville 2015; Guan, Liang et al. 2017; Guan, Yuan et al. 2017; Guo et al. 2017; Hao et al. 2019; He et al. 2015; He et al. 2016; Hinton et al. 2012; Hochreiter and Schmidhuber 1997; Hong and Park 2017; Iandola and Keutzer 2017; Imani et al. 2018; Intel 2020; Ioffe and Szegedy 2015; Jakhar and Hooda 2018; Jia et al. 2014; Krizhevsky, Sutskever, and Hinton 2012; Lawrence and Zhang 2019; Li et al. 2018; Liu et al. 2016; Liu et al. 2020; Luo et al. 2019; University of Minnesota 2013; Mohammadi et al. 2018; Mosavi, Ardabili, and Varkonyi-Koczy 2019; Mozer 2020; Najafabadi et al. 2015; Ndikumana et al. 2020; Njima et al. 2019; O'Shea and Nash 2015; Ovidiu and Peter 2014; Patel and Patel 2016; Redmon 2013; Redmon et al. 2016; Rumelhart, Hinton, and Williams 1986; Samajdar et al. 2018; Samavatian et al. 2020; Shafiee et al. 2016; Sheng et al. 2020; Sodhro et al. 2017; Sodhro et al. 2019; Song et al. 2017; Stergiou et al. 2018; Sze et al. 2017; Szegedy et al.2017; Taigman et al. 2014; Tang 2013; Tang et al. 2017; Umuroglu et al. 2017; Van Gerven and Bohte 2017; Venieris and Bouganis 2016; Venieris, Kouris, and Bouganis 2018; Vipin and Fahmy 2018; Vohra and Fasciani 2019; Wang et al. 2018; Wu et al. 2017; XILINX 2021; Yu et al. 2019; Zagoruyko and Komodakis 2016; Zhang and Kouzani 2020).

REFERENCES

Abadi, Martín, Ashish Agarwal, Paul Barham, Eugene Brevdo, Zhifeng Chen, Craig Citro, Greg S. Corrado, Andy Davis, Jeffrey Dean, and Matthieu Devin. 2016. "Tensorflow: Large-scale machine learning on heterogeneous distributed systems." *arXiv preprint arXiv:1603.04467.*

Albericio, Jorge, Patrick Judd, Tayler Hetherington, Tor Aamodt, Natalie Enright Jerger, and Andreas Moshovos. 2016. "Cnvlutin: Ineffectual-neuron-free deep neural network computing." *ACM SIGARCH Computer Architecture News* 44 (3):1–13.

Al-Garadi, Mohammed Ali, Amr Mohamed, Abdulla Khalid Al-Ali, Xiaojiang Du, Ihsan Ali, and Mohsen Guizani. 2020. "A survey of machine and deep learning methods for internet of things (IoT) security." *IEEE Communications Surveys & Tutorials* 22 (3):1646–1685.

Alwzwazy, Haider A., Hayder M. Albehadili, Younes S. Alwan, and Naz E. Islam. 2016. "Handwritten digit recognition using convolutional neural networks." *International Journal of Innovative Research in Computer and Communication Engineering* 4 (2):1101–1106.

Aly, Mohamed M. Sabry, Mingyu Gao, Gage Hills, Chi-Shuen Lee, Greg Pitner, Max M. Shulaker, Tony F. Wu, Mehdi Asheghi, Jeff Bokor, and Franz Franchetti. 2015. "Energy-efficient abundant-data computing: The N3XT 1,000 x." *Computer* 48 (12):24–33.

Azari, Elham, and SarmaVrudhula. 2020. "ELSA: A throughput-optimized design of an LSTM accelerator for energy-constrained devices." *ACM Transactions on Embedded Computing Systems (TECS)* 19 (1):1–21.

Bajaj, Ronak. 2016. *Exploiting DSP Block Capabilities in FPGA High Level Design Flows.* Nanyang Technological University, Singapore.

Bank-Tavakoli, Erfan, Seyed Abolfazl Ghasemzadeh, Mehdi Kamal, Ali Afzali-Kusha, and Massoud Pedram. 2019. "Polar: A pipelined/overlapped FPGA-based LSTM accelerator." *IEEE Transactions on Very Large Scale Integration (VLSI) Systems* 28 (3):838–842.

Bottou, L. (2010) "Large-scale machine learning with stochastic gradient descent." Lechevallier Y., Saporta G. (eds). *Proceedings of COMPSTAT'2010.* Physica-Verlag HD. https://doi.org/10.1007/978-3-7908-2604-3_16.

Castro-Godínez, Jorge, Deykel Hernández-Araya, Muhammad Shafique, and Jörg Henkel. 2020."Approximate acceleration for CNN-based applications on IoT edge devices." *IEEE 11th Latin American Symposium on Circuits & Systems (LASCAS)* (pp. 1–4), doi: 10.1109/LASCAS45839.2020.9069040.

Chauvin, Yves, and David E. Rumelhart. 1995. *Backpropagation: Theory, Architectures, and Applications.* Psychology Press. https://books.google.co.in/books?hl=en&lr=&id= 3kyC7Woul OQC&oi=fnd&pg=PR3&dq=Backpropagation:+theory,+arch itectures,+and+applications&ots=Ntq7XD6JeH&sig=02SCPwXjwnWM_ JpPPhVh63WMe5E&redir_esc=y#v=onepage&q=Backpropagation%3A%20 theory%2C%20architectures%2C%20and%20applications&f=false.

Chen, Tianqi, Mu Li, Yutian Li, Min Lin, Naiyan Wang, Minjie Wang, Tianjun Xiao, BingXu, Chiyuan Zhang, and Zheng Zhang. 2015. "Mxnet: A flexible and efficient machine learning library for heterogeneous distributed systems." *arXiv preprint arXiv:1512.01274.*

Chen, Tianshi, Zidong Du, Ninghui Sun, Jia Wang, Chengyong Wu, Yunji Chen, and Olivier Temam. 2014. "Diannao: A small-footprint high-throughput accelerator for ubiquitous machine-learning." *ACM SIGARCH Computer Architecture News* 42 (1):269–284.

Chen, Yu-Hsin, Joel Emer, and Vivienne Sze. 2016. "Eyeriss: A spatial architecture for energy-efficient dataflow for convolutional neural networks." *ACM SIGARCH Computer Architecture News* 44 (3):367–379.

Chen, Yu-Hsin, Joel Emer, and Vivienne Sze. 2017. "Using dataflow to optimize energy efficiency of deep neural network accelerators." *IEEE Micro* 37 (3):12–21.

Cisco. 2020. "Cisco Annual Internet Report (2018–2023) White Paper." https://www.cisco.com/c/en/us/solutions/collateral/executive-perspectives/annual-internet-report/white-paper-c11-741490.html (accessed January 21, 2021).

Cuervo, Eduardo, Aruna Balasubramanian, Dae-ki Cho, Alec Wolman, Stefan Saroiu, Ranveer Chandra, and Paramvir Bahl. 2010. "MAUI: making smartphones last longer with code offload." *Proceedings of the 8th International Conference on Mobile Systems, Applications, and Services (MobiSys'10)* (pp. 49–62) Association for Computing Machinery, New York, NY. https://doi.org/10.1145/1814433.1814441.

Deng, Li, and Dong Yu. 2014. "Deep learning: Methods and applications." *Foundations and Trends in Signal Processing* 7 (3–4):197–387.

Elgawi, Osman, and A. M. Mutawa. 2020. "Low power deep-learning architecture for mobile IoT intelligence." IEEE International Conference on Informatics, IoT, and Enabling Technologies (ICIoT) (pp. 43–47) doi: 10.1109/ICIoT48696.2020.9089642.

Fowers, Jeremy, Kalin Ovtcharov, Michael Papamichael, Todd Massengill, Ming Liu, Daniel Lo, Shlomi Alkalay, Michael Haselman, Logan Adams, and Mahdi Ghandi. 2018. "A Configurable Cloud-Scale DNN Processor for Real-Time AI," *ACM/IEEE 45th Annual International Symposium on Computer Architecture (ISCA)* (pp. 1–14), doi: 10.1109/ISCA.2018.00012.

Ge, Fen, Ning Wu, Hao Xiao, Yuanyuan Zhang, and Fang Zhou. 2019. "Compact convolutional neural network accelerator for IoT endpoint SOC." *Electronics* 8 (5):497.

Glorot, Xavier, Antoine Bordes, and Yoshua Bengio. 2011. "Deep sparse rectifier neural networks." *AISTATS*.

Goodfellow, Ian, Yoshua Bengio, and Aaron Courville. 2015. *Deep Learning [Sl]*. MIT Press.

Guan, Yijin, Hao Liang, Ningyi Xu, Wenqiang Wang, Shaoshuai Shi, Xi Chen, Guangyu Sun, Wei Zhang, and Jason Cong. 2017. "FP-DNN: An automated framework for mapping deep neural networks onto FPGAs with RTL-HLS hybrid templates." *2017 IEEE 25th Annual International Symposium on Field-Programmable Custom Computing Machines (FCCM)* (pp. 152–159), IEEE.

Guan, Yijin, Zhihang Yuan, Guangyu Sun, and Jason Cong. 2017."FPGA-based accelerator for long short-term memory recurrent neural networks." *22nd Asia and South Pacific Design Automation Conference (ASP-DAC)* (pp. 629–634), doi: 10.1109/ASPDAC.2017.7858394.

Guo, Kaiyuan, Shulin Zeng, Jincheng Yu, Yu Wang, and Huazhong Yang. 2017. "A survey of FPGA-based neural network accelerator." *arXiv preprint arXiv:1712.08934*.

Hao, Cong, Xiaofan Zhang, Yuhong Li, Sitao Huang, Jinjun Xiong, Kyle Rupnow, Wen-mei Hwu, and Deming Chen. 2019. "FPGA/DNN co-design: An efficient design methodology for IoT intelligence on the edge." *2019 56th ACM/IEEE Design Automation Conference (DAC)* (pp. 1–6), IEEE.

He, Kaiming, Xiangyu Zhang, Shaoqing Ren, and Jian Sun. 2015. "Delving deep into rectifiers: Surpassing human-level performance on imagenet classification." *Proceedings of the IEEE International Conference on Computer Vision* (pp. 1026–1034).

He, Kaiming, Xiangyu Zhang, Shaoqing Ren, and Jian Sun. 2016. "Deep residual learning for image recognition." *Proceedings of the IEEE Conference on Computer Vision and Pattern Recognition* (pp. 770–778).

Hinton, Geoffrey, Li Deng, Dong Yu, George E. Dahl, Abdel-rahman Mohamed, Navdeep Jaitly, Andrew Senior, Vincent Vanhoucke, Patrick Nguyen, and Tara N. Sainath. 2012. "Deep neural networks for acoustic modeling in speech recognition: The shared views of four research groups." *IEEE Signal Processing Magazine* 29 (6):82–97.

Hochreiter, Sepp, and Jürgen Schmidhuber. 1997. "Long short-term memory." *Neural Computation* 9 (8):1735–1780.

Hong, S., and Park, Y. 2017. A FPGA-based neural accelerator for small IoT devices. *2017 International SoC Design Conference (ISOCC)* (pp. 294–295), IEEE.

Iandola, Forrest, and Kurt Keutzer. 2017. *Keynote: Small Neural Nets Are Beautiful: Enabling Embedded Systems with Small Deep-Neural-Network Architectures*. IEEE. https://ieeexplore.ieee.org/abstract/document/8101283 (accessed 2017).

Imani, Mohsen, Mohammad Samragh, Yeseong Kim, Saransh Gupta, Farinaz Koushanfar, and Tajana Rosing. 2018. "Rapidnn: In-memory deep neural network acceleration framework." *arXiv preprint arXiv:1806.05794*.

Intel. 2020. "Intel ARIA SoC development kit."https://www.intel.com/content/www/us/en/programmable/products/boards_and_kits/dev-kits/altera/arria-10-soc-development-kit.html.

Ioffe, Sergey, and Christian Szegedy. 2015. "Batch normalization: Accelerating deep network training by reducing internal covariate shift." *International Conference on Machine Learning* (pp. 448–456), PMLR.

Jakhar, Karan, and Nishtha Hooda. 2018. "Big data deep learning framework using Keras: A case study of Pneumonia prediction." *4th International Conference on Computing Communication and Automation (ICCCA)* (pp. 1–5), IEEE.

Jia, Yangqing, Evan Shelhamer, Jeff Donahue, Sergey Karayev, Jonathan Long, Ross Girshick, Sergio Guadarrama, and Trevor Darrell. 2014. "Caffe: Convolutional architecture for fast feature embedding." *Proceedings of the 22nd ACM International Conference on Multimedia* (pp. 675–678).

Krizhevsky, Alex, Ilya Sutskever, and Geoffrey E. Hinton. 2012. "Imagenet classification with deep convolutional neural networks." *Advances in Neural Information Processing Systems*. https://kr.nvidia.com/content/tesla/pdf/machine-learning/imagenet-classification-with-deep-convolutional-nn.pdf.

Lawrence, Tom, and Li Zhang. 2019. "IoTNet: An efficient and accurate convolutional neural network for IoT devices." *Sensors*19 (24):5541.

Li, Jin, Yinghui Zhang, Xiaofeng Chen, and Yang Xiang. 2018. "Secure attribute-based data sharing for resource-limited users in cloud computing." *Computers & Security*, Elsevier. https://www.sciencedirect.com/science/article/pii/S0167404817301621.

Liu, Wei, Dragomir Anguelov, Dumitru Erhan, Christian Szegedy, Scott Reed, Cheng-Yang Fu, and Alexander C. Berg. 2016."SSD: Single shot multibox detector. *European Conference on Computer Vision* (pp. 21–37), Springer, Cham.

Liu, Zhiqiang, Jingfei Jiang, Guoqing Lei, Kai Chen, Buyue Qin, and Xiaoqiang Zhao. 2020. "A heterogeneous processor design for CNN-based AI applications on IoT devices." *Procedia Computer Science*174:2–8.

Luo, Zhengyi, Austin Small, Liam Dugan, and Stephen Lane. 2019. "Cloud Chaser: Real time deep learning computer vision on low computing power devices." *Eleventh International Conference on Machine Vision (ICMV 2018)* (Vol. 11041, pp. 743–750), SPIE.

Mohammadi, Mehdi, Ala Al-Fuqaha, Sameh Sorour, and Mohsen Guizani. 2018. "Deep learning for IoT big data and streaming analytics: A survey." *IEEE Communications Surveys & Tutorials* 20 (4):2923–2960.

Mosavi, Amir, Sina Ardabili, and Annamaria R. Varkonyi-Koczy. 2019."List of deep learning models." *International Conference on Global Research and Education* (pp. 202–214), Springer, Cham.

Mozer, Todd F. 2020. "Triggering video surveillance using embedded voice, speech, or sound recognition." Google Patents.

Najafabadi, Maryam M., Flavio Villanustre, Taghi M. Khoshgoftaar, Naeem Seliya, Randall Wald, and Edin Muharemagic. 2015. "Deep learning applications and challenges in big data analytics." *Journal of Big Data* 2 (1):1–21.

Ndikumana, Anselme, Nguyen H. Tran, Ki Tae Kim, and Choong Seon Hong. 2020. "Deep learning based caching for self-driving cars in multi-access edge computing." *IEEE Transactions on Intelligent Transportation Systems* 22(5), 2862–2877.

Njima, Wafa, Iness Ahriz, Rafik Zayani, Michel Terre, and Ridha Bouallegue. 2019. "Deep CNN for indoor localization in IoT-sensor systems." *Sensors* 19 (14):3127.

O'Shea, Keiron, and Ryan Nash. 2015. "An introduction to convolutional neural networks." *arXiv preprint arXiv:1511.08458*.

Ovidiu, Vermesan, and Friess Peter. 2014. *Internet of Things: From Research and Innovation to Market Deployment*. (Vol. 29), Aalborg: River Publishers.

Patel, Keyur K., and Sunil M. Patel. 2016. "Internet of things-IOT: Definition, characteristics, architecture, enabling technologies, application & future challenges." *International Journal of Engineering Science and Computing* 6 (5).

Redmon, Joseph. 2013. "Darknet: Open source neural networks in c." http://pjreddie.com/darknet

Redmon, Joseph, Santosh Divvala, Ross Girshick, and Ali Farhadi. 2016. "You only look once: Unified, real-time object detection." *Proceedings of the IEEE Conference on Computer Vision and Pattern Recognition* (pp. 779–788).

Rumelhart, David E., Geoffrey E. Hinton, and Ronald J. Williams. 1986. "Learning representations by back-propagating errors." *Nature* 323 (6088):533–536.

Samajdar, Ananda, Yuhao Zhu, Paul Whatmough, Matthew Mattina, and Tushar Krishna. 2018. "Scale-sim: Systolic CNN accelerator simulator." *arXiv preprint arXiv:1811.02883*.

Samavatian, Mohammad Hossein, Any sBacha, Li Zhou, and Radu Teodorescu. 2020. "RNNFast: An accelerator for recurrent neural networks using domain-wall memory." *ACM Journal on Emerging Technologies in Computing Systems (JETC)* 16 (4):1–27.

Shafiee, Ali, Anirban Nag, Naveen Muralimanohar, Rajeev Balasubramonian, John Paul Strachan, Miao Hu, R. Stanley Williams, and Vivek Srikumar. 2016. "ISAAC: A convolutional neural network accelerator with in-situ analog arithmetic in crossbars." *ACM SIGARCH Computer Architecture News* 44 (3):14–26.

Sheng, Sun, Li Xujing, Liu Min, Yang Bo, and Guo Xiaobing. 2020. "DNN inference acceleration via heterogeneous IoT devices collaboration." *Journal of Computer Research and Development* 57 (4):709.

Sodhro, Ali Hassan, Faisal K. Shaikh, Sandeep Pirbhulal, Mir Muhammad Lodro, and Madad Ali Shah. 2017. "Medical-QoS based telemedicine service selection using analytic hierarchy process." In *Handbook of Large-Scale Distributed Computing in Smart Healthcare*. Springer. https://link.springer.com/chapter/10.1007/978-3-319-58280-1_21 (accessed January 25, 2021).

Sodhro, Ali Hassan, Zongwei Luo, Gul Hassan Sodhro, Muhammad Muzamal, Joel J. P. C. Rodrigues, and Victor Hugo C. De Albuquerque. 2019. "Artificial intelligence based QoS optimization for multimedia communication in IoV systems." *Future Generation Computer Systems* 95:667–680.

Song, Linghao, Xuehai Qian, Hai Li, and Yiran Chen. 2017. "Pipelayer: A pipelined reram-based accelerator for deep learning." *2017 IEEE International Symposium on High Performance Computer Architecture (HPCA)* (pp. 541–552), IEEE.

Stergiou, Christos, Kostas E. Psannis, Byung-Gyu Kim, and Brij Gupta. 2018. "Secure integration of IoT and cloud computing." *Future Generation Computer Systems* 78:964–975.

Sze, Vivienne, Yu-Hsin Chen, Tien-Ju Yang, and Joel S. Emer. 2017. "Efficient processing of deep neural networks: A tutorial and survey." *Proceedings of the IEEE* 105 (12):2295–2329.

Szegedy, Christian, Sergey Ioffe, Vincent Vanhoucke, and Alexander Alemi. 2017."Inception-v4, inception-resnet and the impact of residual connections on learning." *Thirty-first AAAI Conference on Artificial Intelligence*.

Taigman, Yaniv, Ming Yang, Marc'Aurelio Ranzato, and Lior Wolf. 2014. "Deepface: Closing the gap to human-level performance in face verification." *Proceedings of the IEEE Conference on Computer Vision and Pattern Recognition* (pp. 1701–1708).

Tang, Jie, Dawei Sun, Shaoshan Liu, and Jean-Luc Gaudiot. 2017. "Enabling deep learning on IoT devices." In *Computer*. IEEE. https://ieeexplore.ieee.org/abstract/document/8057306.

Tang, Yichuan. 2013. "Deep learning using linear support vector machines." *arXiv preprint arXiv:1306.0239*.

Umuroglu, Yaman, Nicholas J. Fraser, Giulio Gambardella, Michaela Blott, Philip Leong, Magnus Jahre, and Kees Vissers. 2017. "Finn: A framework for fast, scalable binarized neural network inference." *Proceedings of the 2017 ACM/SIGDA International Symposium on Field-Programmable Gate Arrays* (pp. 65–74).

University of Minnesota. 2013. "A modular approach to Spintronics" (accessed February 6, 2021). http://cspin.umn.edu/resources/modular_approach.html.

Van Gerven, Marcel, and Sander Bohte. 2017. "Artificial neural networks as models of neural information processing." *Frontiers in Computational Neuroscience* 11:114.

Venieris, Stylianos I., Alexandros Kouris, and Christos-Savvas Bouganis. 2018. "Toolflows for mapping convolutional neural networks on FPGAs: A survey and future directions." *arXiv preprint arXiv:1803.05900*.

Venieris, Stylianos I., and Christos-Savvas Bouganis. 2016. "fpgaConvNet: A framework for mapping convolutional neural networks on FPGAs." *2016 IEEE 24th Annual International Symposium on Field-Programmable Custom Computing Machines (FCCM)* (pp. 40–47), IEEE.

Véstias, Mário. 2020. "Processing systems for deep learning inference on edge devices." In *Convergence of Artificial Intelligence and the Internet of Things*, 213–240. Springer.

Vipin, Kizheppatt, and Suhaib A.Fahmy. 2018. "FPGA dynamic and partial reconfiguration: A survey of architectures, methods, and applications." *ACM Computing Surveys (CSUR)* 51 (4):1–39.

Vohra, Manohar, and Stefano Fasciani. 2019."PYNQ-Torch: A framework to develop PyTorch accelerators on the PYNQ platform."

Wang, Junsong, Qiuwen Lou, Xiaofan Zhang, Chao Zhu, Yonghua Lin, and Deming Chen. 2018. "Design flow of accelerating hybrid extremely low bit-width neural network in embedded FPGA."

Wu, Bichen, Forrest Iandola, Peter H. Jin, and Kurt Keutzer. 2017. "Squeezedet: Unified, small, low power fully convolutional neural networks for real-time object detection for autonomous driving." *Proceedings of the IEEE Conference on Computer Vision and Pattern Recognition Workshops* (pp. 129–137).

XILINX. 2021. https://www.xilinx.com/products/%20acceleration-solutions/1-zz0jo0.html.

Y. Guan et al., "FP-DNN: An Automated Framework for Mapping Deep Neural Networks onto FPGAs with RTL-HLS Hybrid Templates," 2017 IEEE 25th Annual International Symposium on Field-Programmable Custom Computing Machines (FCCM), 2017, pp. 152–159, doi: 10.1109/FCCM.2017.25.

Yu, Yong, Xiaosheng Si, Changhua Hu, and Jianxun Zhang. 2019. "A review of recurrent neural networks: LSTM cells and network architectures." *Neural Computation* 31 (7):1235–1270.

Zagoruyko, Sergey, and Nikos Komodakis. 2016. "Wide residual networks. arXiv." *arXiv preprint arXiv:1605.07146*.

Zhang, Zhichao, and Abbas Z. Kouzani. 2020. "Implementation of DNNs on IoT devices." *Neural Computing and Applications* 32 (5):1327–1356.

Zhang, Chen, Peng Li, Guangyu Sun, Yijin Guan, Bingjun Xiao, and Jason Cong. 2015. "OPTIMIZING FPGA-BASED ACCELERATOR DESIGN FOR DEEP CONVOLUTIONAL NEURAL NETWORKS." *Proceedings of the 2015 ACM/SIGDA international symposium on field-programmable gate arrays* (pp. 161–170).

Part III

Communication Technologies and Trends

7 Communication Technologies for M2M and IoT Domain

Manoj Kumar
STMicroelectronics
Noida, India

Sushil Kumar
Telecommunication Engineering Center (TEC)
New Delhi, India

CONTENTS

7.1	Introduction	134
7.2	M2M Communication	134
7.3	Internet of Things	135
7.4	Projections on Connected Devices	136
7.5	Communication Technologies for M2M/IoT	137
7.6	Cellular Technologies	137
	7.6.1 Cellular LPWAN technologies	138
7.7	Non-Cellular Wireless Technologies	139
	7.7.1 Non Cellular / Non 3GPP LPWAN Technologies (Low-Power Wide Area Networks)	140
	7.7.1.1 LoRa	141
	7.7.1.2 Symphony Link	143
	7.7.1.3 Sigfox	143
	7.7.1.4 Ingenu RPMA	145
	7.7.2 Short-Range Communication Technologies	146
	7.7.2.1 RFID	146
	7.7.2.2 NFC	147
	7.7.2.3 Bluetooth and Bluetooth Low Energy	148
	7.7.2.4 ZigBee	150
	7.7.2.5 6LoWPAN	151
	7.7.2.6 Z-Wave	152
	7.7.2.7 EnOcean	153
	7.7.2.8 ANT/ANT+	153
7.8	IoT Hubs and Gateways	153
7.9	Topologies: Mesh vs. Star	155

DOI: 10.1201/9781003122357-10

7.10 IoT Communication Protocols 155
7.11 Conclusion 157
References 157

7.1 INTRODUCTION

The installed base of Internet of things (IoT) and machine-to-machine (M2M) communication nodes continues to increase year over year. Communication among M2M/IoT nodes can broadly be divided between cellular and non-cellular. There are vast variety of communicating devices in security, automation, ATMs, POS terminals, remote displays, and healthcare equipment etc. that continue to transmit data either on demand, periodically, or need based. These can be standalone devices communicating with a server or could be part of a communication network where these could be transmitting to neighbors which retransmit it as per the specific protocols being used in a given environment. M2M / IoT has a long list of application areas, e.g., intelligent transportation systems, logistics and supply chain management, smart metering, e-healthcare, surveillance and security, smart cities, and home automation (Mehmood et al., 2015).

When each of the device can be reached by a cellular network, the network management is handled as per the established cellular standards. When devices are having other communication interfaces, there is a need of either private networks or local gateways that in turn can make each of the device available on the cloud or a central server. Key point in both these cases is addressability.

In this chapter, we will discuss about various cellular and non-cellular technologies being used in M2M and IoT.

7.2 M2M COMMUNICATION

M2M stands for machine-to-machine communication. It is a concept where two or more than two machines communicate with each other using a wired or wireless medium without human intervention. M2M communication describes a communication style in which two or more entities such as devices/machines communicate with each other autonomously (Mehmood et al., 2015). M2M communication finds use in several applications like manufacturing, security, facility management, attendance systems, and point of sale (PoS) terminals. The term M2M in IoT describes the autonomous exchange of information among numerous devices inter-connected with each other.

It is important to note that M2M traffic differs significantly compared with the conventional cellular or mobile traffic. The mobile traffic is primarily downstream while M2M is primarily upstream. M2M devices compete for the data exchange with the cellular traffic. This has been the reason for LTE and MTC requirements being addressed by 3GPP (Mehmood et al., 2015).

M2M architecture is shown in Figure 7.1. The M2M devices are responsible for the collection and autonomous transmission of sensory or other type of data. The M2M devices are usually connected to small local networks called subnets for the

Communication Technologies for M2M and IoT Domain

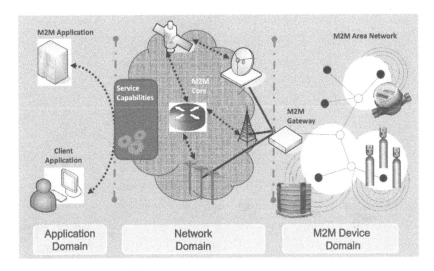

FIGURE 7.1 M2M architecture. (From TEC, 2017b.)

transmission or reception of data to or from the M2M application domains. A gateway collects the information of all the M2M devices in a designated area and sends out this information over a wired or wireless network to the M2M application that can be at a cloud or other central system (Mehmood et al., 2015).

The enabling technologies for M2M communication are sensor networks, RFID, mobile Internet, IPv4/IPv6, etc. While low-power wireless communication technologies such as Wi-Fi, ZigBee, 6LoWPAN, Bluetooth Low Energy (BLE), Z-Wave etc. may be used to connect the devices with the M2M gateway and to communicate to each other in a local network, long range technologies like 4th or 5th generation cellular, fixed line broadband/FTTH may be used for connecting M2M gateway to the server. In cases where traditional cellular network may not be available or not appropriate for the application, low-power wide area network (LPWAN) technologies as mentioned in Figure 7.4, may be used for transmitting very small data over a long range.

7.3 INTERNET OF THINGS

ITU-T has defined IoT, as a global infrastructure for the information society, enabling advanced services by interconnecting (physical and virtual) things based on existing and evolving interoperable information and communication technologies in its Recommendation ITU-T Y.2060 (06/2012) (ITU-T, 2012).

The IoT is extension of M2M communication technology, where the devices can not only communicate to other similar devices but additionally, they can also communicate over the Internet. This makes it possible to have a large network of heterogeneous devices communicating securely over standard TCP/IP protocols. Being a part of IoT network does not necessarily require huge upgrades to the devices, they can keep on using simpler non-IP protocols specific to the use case and connect to Internet through an IP-enabled gateway. The biggest advantage of IoT over traditional M2M is the ability to leverage virtually unlimited resources provided by cloud computing infrastructure.

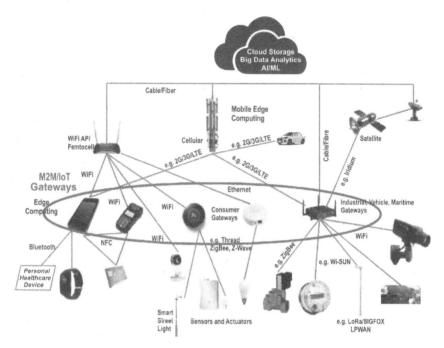

M2M/IoT Architecture for Connecting Smart Devices

FIGURE 7.2 The IoT network. (Adopted from TEC, 2017a.)

The IoT ecosystem includes devices (sensors and actuators), gateways, communication technologies, security protocols, and user interfaces (web, mobile, HMI). Governance and process management also play a critical role in an IoT ecosystem. IoT systems tend to generate a huge amount of data. Big data analytics may be used to create intelligence, which may be further used for various operational and planning activities. A typical IoT network along with communication technologies and gateways is shown in Figure 7.2.

The enabling technologies for M2M/IoT are sensor networks, RFID, mobile Internet, location-based services (LBS), augmented reality (AR), artificial intelligence, wired and wireless communication network, IPv4/IPv6, etc.

7.4 PROJECTIONS ON CONNECTED DEVICES

Looking at the installed base of devices supporting various connectivity technologies, Wi-Fi clearly dominates with projected usage of 27% by 2026, followed by Bluetooth with 23% projected market share in 2026. Figure 7.3 shows the IoT connections by technology.

Main IoT applications covered in the chart are: home security and automation, asset tracking, home appliances, OEM telematics, smart meters, aftermarket telematics, and inventory management.

IoT	2020	2026	CAGR
Wide-area IoT	1.9	6.3	22%
Cellular IoT	1.7	5.9	23%
Short-range IoT	10.7	20.6	12%
Total	12.6	26.9	13%

FIGURE 7.3 Projections for IoT connections in billions.[1]

As per the recent projections by Ericsson, there may be around 26.9 billion connected devices across the globe by 2026.[2] It is interesting to observe that past estimates by various forecasting firms have been more ambitious than the actual figure, yet the growth in the devices remains impressive.

7.5 COMMUNICATION TECHNOLOGIES FOR M2M/IoT

M2M/IoT will have a large number of communication technologies such as RFID, NFC, BLE working in very short range, low-power wireless, cellular (2G, 3G, 4G, 5G), low-power wide area network (LPWAN) technologies, fixed line broadband, power line communication, etc.

LPWAN technologies have been developed to carry a very small data to a large distance and consume very low power. It covers 2–3 km in city (dense) areas and 12–15 km in rural (open) areas. Battery life is expected to be around 8–10 years, and depends mainly on the use case requirement. Cellular LPWAN (EC-GSM, NB-IoT, LTE-MTC) are based on 3GPP Release 13 and beyond.[3] Cellular operators can enable LPWAN services in the existing GSM/LTE networks by upgrading the software. Trials have been done and the commercial offerings are also available in a number of countries namely South Korea, Europe, USA, etc. Non-cellular/non-3GPP LPWAN technologies such as LoRa, Sigfox, etc. (Figure 7.4) have been described in section 7.7.1.

Use cases are in the areas of smart metering, smart farming (transmitting soil testing data), smart bin, transmitting pollution sensor data, sending alerts, etc.

7.6 CELLULAR TECHNOLOGIES

Figure 7.5 provides a good overview of various communication technologies relevant to M2M and IoT. Clearly, there are more as we will discuss later in this chapter but it is a good starting point for our discussion. As the normal SIM card is not suitable for harsh conditions of vehicles like vibrations, temperature, and humidity. GSMA

---

1. IoT Connections Outlook, https://www.ericsson.com/en/mobility-report/dataforecasts/iot-connections-outlook, Accessed 13 Feb 2021.
2. IoT Connections Outlook, https://www.ericsson.com/en/mobility-report/dataforecasts/iot-connections-outlook, Accessed 13 Feb 2021.
3. Release 13, https://www.3gpp.org/release-13, Accessed 14 Feb 2021

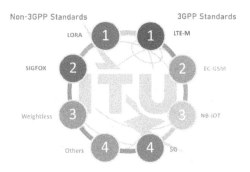

FIGURE 7.4 Long-range technologies. (From Tabbane, 2018.)

has created specifications for embedded M2M SIM, with over-the-air (OTA) provisioning. Temperature variation range is from −40 degree to +125 degree Celsius. Embedded SIM technology offers big opportunities for auto manufacturers as the lifecycle of an eSIM is around 10–12 years.

7.6.1 Cellular LPWAN Technologies

3GPP released the specifications in its release 13 and beyond for LTE-MTC, NB-IoT and EC-GCM as a low power wide area network technologies, suitable for M2M/IoT communication.

LTE-M for (M2M) communications works within the normal infrastructure of LTE networks and does not need additional investments by the operators. LTE-M receivers only need to process 1.4 MHz of the channel instead of 20 MHz, which makes it become much simpler than mobile phone receivers (Qin et al., 2019).

NB-IoT coexists with GSM or LTE existing licensed bands. This results into two types of NB-IoT technologies—one based on GSM and another based on LTE.

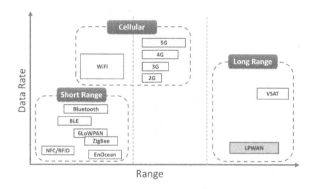

FIGURE 7.5 Data rate vs. range for various M2M/IoT communication technologies. (From Mekki et al., 2018.)

Communication Technologies for M2M and IoT Domain

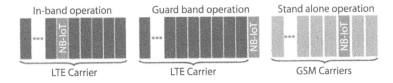

FIGURE 7.6 Operation modes for NB-IoT. (From Mekki et al., 2018.)

For the case of GSM, it occupies a bandwidth of 200 KHz. It can operate in standalone, guard band, or in-band modes as shown in Figure 7.6. NB-IoT reuses various principles and building blocks of LTE physical and higher protocol layers. It reduces LTE protocol functionalities to the minimum and enhances them as needed for IoT applications. The high compatibility with GSM and LTE provides NB-IoT with great potential for worldwide deployment. The countries where GSM networks have been phased out can make use of LTE and those that still have GSM networks, they can use their existing infrastructure of 2G/3G. Besides, there are country-specific preferences as well. For example, China and India adopting NB-IoT whereas North America LTE-MTC.

NB-IoT allows connectivity of more than 100K devices per cell and it could further be increased by using multiple NB-IoT carriers. It employs QPSK modulation, frequency division multiple access (FDMA) in uplink, and orthogonal FDMA (OFDMA) in downlink. The maximum throughput rate is 200 kbps in downlink and 20 kbps in the uplink. NB-IoT and LTE-M technologies can achieve 10 years of battery lifetime by transmitting on average 200 bytes per day (Mekki et al., 2018).

NB-IoT focuses mainly upon MTC class of devices that are installed at places far from usual reach. Thus, it is not suitable for rural or suburban regions that do not have 4G coverage. NB-IoT can be deployed by reusing and upgrading the existing cellular network but its deployments are restricted to the area supported by cellular network. NB-IoT is directed toward applications that require high QoS and low latency (Sinha et al., 2017).

7.7 NON-CELLULAR WIRELESS TECHNOLOGIES

Non-cellular wireless technologies (also wired like power line communication and others but are not discussed in this chapter) have their own advantages and considerations. The cellular communication offers seamless connectivity and extension of existing practice. Non-cellular requires either a network ownership or a subscription. LPWAN technologies, primarily Sigfox and LoRA, are seeing wider adoption. LoRA allows or requires a private network to be installed, whereas Sigfox provides its own subscription model to connect its nodes.

Within non-cellular also there is a further segmentation possible. Low-power and short-range wireless technologies such as Bluetooth, ZigBee have been developed as last mile connectivity to connect end-devices to gateways. On the other hand, radio technologies like Sigfox and LoRa have been developed for end-to-end connections for applications with limited demands on throughput.

Most IoT devices operate in industrial, scientific, and medical (ISM) band of the spectrum. Many countries have de-licensed spectrum in various slots in sub-GHz

TABLE 7.1
Sub-GHz ISM Bands in Various Countries

Country	Frequency Band
North America, Mexico, and South America	433.075–434.775 MHz and 902–928 MHz
Africa and Middle Eastern countries	433.05–434.79 MHz and 863–870 MHz
Europe	433.05–434.79 MHz, 863–870 MHz, 870–876 MHz, and 915–921 MHz
Japan	426–430 MHz and 920–928 MHz
Australia/New Zealand	915–928 MHz
India	433–434 MHz and 865–867 MHz
China	470–510 MHz and 920.5–924.5 MHz
Singapore	866–869 MHz and 920–925 MHz
Hong Kong/Thailand/Vietnam	920–925 MHz
Brazil	902–907.5 MHz and 915–928 MHz
Philippines	915–918 MHz
Malaysia	919–923 MHz

Source: TEC (2017a).

band which can enable the M2M/IoT services. Table 7.1 gives an overview of the spectrum band country wise.

The non-cellular M2M/IoT technologies are divided in the following categories and discussed in this order later:

- LPWAN technologies (up to a few kilometers)
- Short-range communication technologies (few centimeters to few hundred meters)

7.7.1 Non Cellular / Non 3GPP LPWAN Technologies (Low-Power Wide Area Networks)

Non cellular LPWAN technologies are designed to transmit very low amount of data (such as meter readings, sensor data from pollution devices, etc.) to large distances. LoRa and Sigfox are examples of such technologies. Their range can stretch beyond 10 km in open air and devices can have a battery life of up to 10 years. These technologies use unlicensed frequency bands worldwide in sub-GHz band. Such high range is achieved by use of deeper modulation schemes, redundant encoding schemes, and high receiver sensitivity. Sigfox provides the end-to-end solution including the communication technology as well as the network infrastructure; on the other hand, LoRa only provides the communication technology.

Several such technologies exist today, such as LoRa, Sigfox, RPMA, Telensa, and Weightless. We will discuss some of these briefly in this section.

7.7.1.1 LoRa

LoRa stands for Long Range (LoRa® Alliance, 2015). LoRa® is the physical layer utilizing specific wireless modulation to create the long-range communication link. It uses chirp spread spectrum modulation, which has low power characteristics like FSK modulation but significantly increases the range. LoRa® requires very good receiver sensitivity.

The advantage of LoRa® is in the technology's long-range capability. A single gateway or base station can cover entire cities or hundreds of square kilometers. With a minimal amount of infrastructure, entire country can easily be covered. The LoRa devices are connected in a star topology network, i.e., each device connects to the gateway, unlike in a mesh.

LoRa is designed to work on unlicensed frequency bands such as 433 MHz, 868 MHz, or 915 MHz, depending on the geographical location and the corresponding regulations (Peña Queralta et al., 2019).

Key features of LoRa technology are:

- Long range: 2–4 km in city/congested areas and 12–15 km in rural or open areas.
- Low power: 5–10 years is the expected battery lifetime.
- Low cost: network deployment cost is very low.
- Secure: with embedded end-to-end AES-128 encryption of data.
- Geolocation: enables indoor/outdoor localization without GPS with an accuracy of around 100 m.

The LoRa® system has three components:

1. LoRa® end-devices.
2. LoRa® gateways: concentrators connecting end-devices to the LoRa® network server.
3. LoRa® network server: the network server that controls the whole network.

The link layer of LoRa LPWAN networks is referred to as LoRaWAN. The MAC layer that operates on top of the LoRa PHY layer is defined in LoRaWAN specifications (Yousuf et al., 2019). LoRaWAN defines the communication protocol and system architecture for the network. LoRaWAN was standardized by LoRa-Alliance in 2015 (Mekki et al., 2018). LoRaWAN defines multiple communication classes for addressing the different latency in IoT applications (Figure 7.7). The LoRaWAN end-devices are classified as Class A, Class B, and Class C and described below (Mekki et al., 2018):

- *Class A* (bidirectional end-devices): End-devices allow bidirectional communication where each end-device's uplink transmission is followed by two short windows for receiving downlink messages. The uplink transmission time is scheduled by the end-device based on its own communication needs. This class is the lowest power end-device system for IoT applications that only require short downlink communication after the end-device has

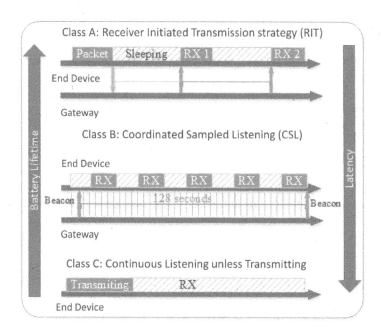

FIGURE 7.7 LoRaWAN end-devices classification. (From Ayoub et al., 2019.)

sent an uplink message. Downlink transmission at any other time will have to wait until the next uplink message of the end-device.
- *Class B* (bidirectional end-devices with scheduled receive slots/sampled listening): In addition to the random receive windows of class A, the class B devices open extra receive windows at scheduled times. In order to open receive window at the scheduled time, end-devices receive a time synchronized Beacon from the base station. This allows the server to know when the end-device is listening.
- *Class C* (bidirectional end-devices with maximal receive slots/continuous listening): End-devices have nearly continuously open receive windows. This class consumes excessive energy and is defined for IoT applications with continuous energy power resources.

The network is existing in a star topology, where the end-devices are connected via single-hop LoRa® communication to one or more gateways. The gateways relay messages between end-devices and network server. Network nodes are not associated with a specific gateway. Data from a node is received by multiple gateways. Each gateway will forward the received packet from the end-node to the cloud-based network server via cellular/Ethernet backhaul. The network server will filter redundant received packets, perform security checks, schedule acknowledgments through the optimal gateway, and perform adaptive data rate, etc. The gateways are transparent to the end-devices, which are logically connected directly to the network server (Figure 7.8) (LoRa® Alliance, 2015).

Communication Technologies for M2M and IoT Domain

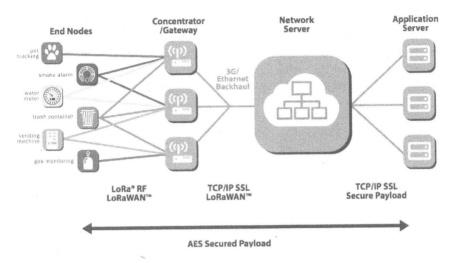

FIGURE 7.8 LoRaWAN protocol. (From LoRa® Alliance, 2015.)

LoRa has been widely deployed in logistics tracking, asset tracking, smart agriculture, intelligent buildings, factories, facility management, healthcare, and airport management.

7.7.1.2 Symphony Link

LoRaWAN is the most popular network protocol that operates on top of LoRa. However, other open source and proprietary solutions have been designed to tackle the limitations of LoRaWAN. Symphony Link, developed by Link Labs, overcomes some of the limitations of LoRaWAN. Symphony Link puts a stronger focus on industrial applications and provides a solution to eliminate the duty cycle limits by using frequency hopping and different available frequencies. It also provides built-in support for repeaters to increase the range of a single gateway without trading off total bandwidth. Another strong point of Symphony Link is that it allows one-to-many communication from a gateway to end-nodes. This simplifies and increases the speed of over-the-air updates or broadcast messages. In terms of open access, the LoRaWAN is an open standard while Symphony Link is proprietary (Peña Queralta et al., 2019).

7.7.1.3 Sigfox

Sigfox technology is a part of the LPWAN family of technologies, employed mainly for the development of the IoT networks, where the volume of data sent is low, the operating range is higher (reaching tens of km), and the current consumption is very low (in the order of mA or tens of mA per transmission). The star-based topology (one-hop star topology) is employed, and end-devices are directly connected to a base station. The architecture of the network is simplified and the centralization and sending of information to the Internet is done within a well-defined frame at the level of one physical equipment (e.g., Lavric et al., 2019).

Sigfox uses its patented ultra narrow band (UNB) technologies and deploys its proprietary base stations in the unlicensed sub-GHz ISM bands. The end-devices connect to these base stations using BPSK modulation in an ultra-narrow band of 100 Hz at a maximum data rate of 100 bps. The number of messages over the uplink is limited to 140 messages per day. The maximum payload length for each uplink message is 12 bytes. The number of messages over the downlink is limited to 4 messages per day with payload length for each downlink message limited to 8 bytes (Lavric et al., 2019; Mekki et al., 2018).

In Sigfox system, signal processing techniques provide high link budget and effective protection against interference. Sigfox uses an UNB communications radio which allows an increased transmission range while spending a limited amount of energy. UNB also allows a larger number of devices to coexist in each cell. Sigfox system is particularly well adapted for low throughput IoT traffic. Sigfox LPWAN autonomous battery-operated devices send only a few bytes per day while being on a single battery for up to 10–15 years.

The Sigfox solution architecture consists of a single core network, which allows global connectivity. The core network elements are the service center (SC) and the registration authority (RA). The SC oversees the data connectivity between the base station (BS) and the global Internet, as well as the control and management of the BSs and end points. The RA oversees the end point network access authorization. The radio access network is comprised of several base stations (BS) connected directly to the SC. The devices or end points (EPs) communicate with base stations. The Sigfox solution connects to a network application (NA) using a restful architecture (Figure 7.9).

Sigfox network is provided as a service. The coverage is present in several countries already, 72 countries at the time of this writing.[4]

The radio interface is optimized for uplink transmissions, which are asynchronous. Downlink communications are achieved by querying the network for existing data from the device. A device willing to receive downlink messages opens a fixed window for reception after sending an uplink transmission. The Sigfox network transmits the downlink message for a given device during the reception window. The Sigfox service center selects the BS for transmitting the corresponding downlink message. Fewer downlink messages are allowed compared to uplinks due to the regulatory constraints on the ISM bands (Figure 7.10).

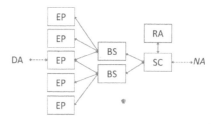

FIGURE 7.9 Sigfox system architecture. (From TEC, 2017a.)

---

4. www.sigfox.com

Communication Technologies for M2M and IoT Domain

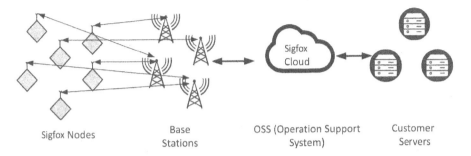

Sigfox Nodes | Base Stations | OSS (Operation Support System) | Customer Servers

FIGURE 7.10 Sigfox architecture. (From Lavric et al., 2019.)

The packet received by the gateway module is sent to Sigfox Cloud using an IP-based network. The architecture uses the Sigfox Cloud which is an interface between the Sigfox customers and the Sigfox partners. The Cloud offers services for map predictions related to coverage and can provide management of network devices of the user accounts. As per the Sigfox technology documentation, the Sigfox Cloud is an OSS (Operation Support System) (Lavric et al., 2019).

7.7.1.4 Ingenu RPMA

Ingenu RPMA is a technology having advanced modulation techniques and was designed with a focus on minimizing the total cost of ownership while increasing the range of a base station and link capacity of LoRa and Sigfox. *Random Phase Multiple Access* (RPMA) is a technology patented in 2010 by Ingenu. On top of it, Ingenu has developed LPWAN technology that allows much higher link capacity than LoRa or Sigfox. It works on the 2.4 GHz ISM band, in contrast with most LPWAN technologies using sub-gigahertz frequencies. This has the benefit of being equivalently regulated over the world. However, the 2.4 GHz is widely used by many other technologies, including Wi-Fi and Bluetooth, and therefore interference is more probable as the spectrum is more congested.

RPMA is based on the direct-sequence spread spectrum (DSSS) modulation technique. Communication is two-way, and devices perform scanning in the background with handover so that the best access point is chosen for each transmission. Its downlink capacity is much larger than in the case of *Symphony Link* due to an adaptive spreading factor methodology. One of the key advantages of RPMA over LoRa and Sigfox is the network capacity. Ingenu claims that a single gateway can handle up to 2 million devices per access point. While LoRa receivers can demodulate signals with different bandwidths or different spreading factors simultaneously, RPMA supports parallel demodulation of up to 1200 signals on the same frequency. As with Symphony Link, Ingenu requires all gateways in the same network to be synchronized, so that end-devices are aligned in time with them. RPMA has a higher link budget, longer range in open space, and the 2.4 GHz band is regulated consisted through the globe, as compared to sub-gigahertz bands. However, a higher frequency also means that penetration through most materials is less effective.

This translates into less range in dense urban areas or large indoor facilities (Peña Queralta et al., 2019).

Ingenu RPMA offers the highest data rate when compared to most LPWAN technologies, though the real transmission speed will adapt according to channel conditions.

7.7.2 Short-Range Communication Technologies

7.7.2.1 RFID

RFID is a term coined for short-range radio technology used to communicate mainly digital information between a stationary location and a movable object or between movable objects. A variety of radio frequencies and techniques are used in RFID systems. RFID is generally characterized by use of simple devices on one end of the link and more complex devices on the other end of the link. The simple devices or tags are small and inexpensive, can be deployed economically in very large numbers, are attached to the objects to be managed and operate automatically. The readers that are used to read the information from tags are more complex and usually connected to a computer or Internet.

A radio-frequency identification system uses tags, attached to the objects to be identified. Two-way radio transmitter-receivers called readers send a signal to the tag and read its response. The readers transmit the read data to a computer system running RFID software (Landt, 2005).

RFID tags can be either passive, active, or battery-assisted passive. An active tag has an on-board battery and periodically transmits its ID signal. A battery-assisted passive has a small battery on board and is activated when in the presence of a RFID reader. Passive RFID tags have the advantages of long life and being small enough to fit into a practical adhesive label. Hence, passive RFID tags are used for many applications.

A passive RFID tag consists of mainly three parts: an antenna, a semiconductor chip attached to the antenna, and some encapsulation to protect the tag from the environment. The passive RFID tags don't carry any powered device and become active only upon exposure to external energy. The RFID reader activates and communicates with the tag (Ahuja & Potti, 2010).

The salient characteristics of this technology are:

- Standard: ISO/IEC 14543-3-10
- Frequency: 120–150 kHz (LF), 13.56 MHz (HF), 433 MHz (UHF), 865–868 MHz (Europe)
- 902–928 MHz (North America) UHF, 2450–5800 MHz (microwave), 3.1–10 GHz (microwave)
- Range: 10 cm to 200 m

The most visible usage of RFID technologies is in tolling on highways. Animal tracking, asset tracking are the other key application areas for RFID technology.

7.7.2.2 NFC

NFC stands for near field communication. It is based on the RFID principles. NFC works at 13.56 MHz and the data rate is 106, 212, and 424 kbps (Coskun et al., 2013). NFC is designed for use by devices within proximity to each other with distances less than 10 cm. NFC is used to share small amount of data, e.g., users can share Bluetooth or Wi-Fi link set-up parameters or exchange data such as virtual business cards. Payments and payment terminals used for contactless payment are based on NFC technology.

NFC finds multiple applications. It can be used for firmware upgrade over the air also in embedded applications. This can be useful in scenarios where a device may not be connected to Internet. A home appliance enabled with NFC could be probed by a service technician and where necessary, field upgrade could happen. Another use case can be in metering. A handheld terminal can be used to get the latest meter reading each billing cycle. All meter parameters like its unique identity, tampering information and consumption can be transferred to NFC handheld device.

There are two types of NFC devices—active and passive devices. An active device is the one that generates the RF field and a passive is the one that only responds. For NFC communication to take place either one or both the devices can be active devices. Active devices need power to operate and while communicating with the other device, they can also deliver a small amount of power through inductive coupling. An active NFC device such as a smartphone can not only read NFC tags but can also exchange information with other compatible devices and even alter the information on the NFC tag if authorized to make such changes.

Tags are the passive devices and work on the principle of inductive coupling when the tag is brought near a transceiver. There are three modes of data transfer in NFC:

- *Peer-to-peer communication:* This mode enables two NFC-enabled devices to communicate with each other to exchange information. Peer-to-peer mode is standardized on the ISO/IEC 18092 standard and is based on NFC Forum's Logical Link Control Protocol Specification.
- *Card emulation:* Card emulation mode enables NFC-enabled (active) devices to act like smart cards, allowing users to perform transactions such as purchases, ticketing, and transit access control using their smartphones.
- *Reader/writer:* Reader/writer mode enables NFC-enabled devices to read information stored on NFC tags.

NFC is useful not just because of the underlying communication technology but because of the stack built on top of it including various standards and abstractions layer on top of the core NFC functionality. NFC is one of the few communication technologies which are adopted by the payment industry. NFC technology finds use in automated fare collection system, ID cards, access control, toll collection, road tolling, smart parking systems, and digital payments.

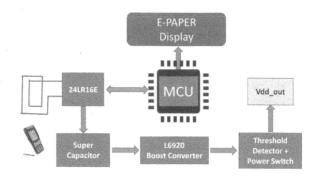

FIGURE 7.11 E-paper display using NFC. (From www.st.com.)

As an interesting application of NFC,[5] e-paper scheme for electronic shelf label is shown in Figure 7.11. Electronic shelf label (ESL) is a system where the price of a commodity is displayed and updated electronically. These are typically reusable modules with a low profile, and of credit card size. These display systems are attached mechanically to the racks holding the article. These systems mostly use liquid crystal display (LCD) or OLEDs to show the applicable product price to the customer.

This type of solution can find many interesting applications ranging from labels in super markets to book stores to hospitals and other places where smart displays are required. Some other variants may be smart ID badges, status displays, electronic patient records, etc.

7.7.2.3 Bluetooth and Bluetooth Low Energy

The Bluetooth Classic radio, also referred to as Bluetooth Basic Rate/Enhanced Data Rate (BR/EDR), is designed for low-power operation and leverages Adaptive Frequency Hopping approach, transmitting data over 79 channels. The Bluetooth BR/EDR radio includes multiple PHY options that support data rates from 1 Mb/s to 3 Mb/s and supports multiple power levels, from 1 mW to 100 mW, multiple security options, and a point-to-point network topology. Bluetooth Special Interest Group (SIG) has several thousand member organizations in its fold.

Bluetooth Low Energy (BLE) devices operate in the same unlicensed 2.4 GHz ISM (Industrial Scientific Medical) band as classic Bluetooth. BLE was introduced as an extension to Bluetooth 4.0 specifications as a low-power solution for control and monitoring applications (Gomez et al., 2012).

Bluetooth LE radio utilizes Frequency-Hopping Spread Spectrum (FHSS) scheme to avoid interference with technologies like Wi-Fi operating in the same frequency range. Unlike the classical Bluetooth, which uses 79: 1-MHz-wide channels, BLE uses 40: 2-MHz-wide channels. Three of these channels, which are located between commonly used wireless local area network channels, are used for advertising and service discovery, and are called advertising channels. The remaining 37 data channels are used to transfer the data. When two Bluetooth devices are connected, they operate their transmitter and receiver in synchronization.

---

5. Energy Harvested Electronic Shelf Label, Technical article by STMicroelectronics, www.st.com

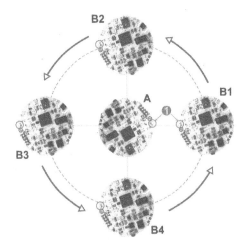

FIGURE 7.12 BLE-based proximity sensing. (From Singhal & Tiwari, 2020.)

BLE has rapidly advanced in recent times. Key reason in its adoption has been the support available in the mobile phones. Any BLE network can be controlled by the APPs on a smartphone that makes it an incredible choice and the key differentiator for applications in personal electronics and smart home domains (TEC, 2017b). It is this very feature that makes many healthcare devices more meaningful. For example, a MEMS-based device can be used for indoor navigation with BLE enablement for localization inside the buildings where GPS signal is not present (Satan, 2018).

A social distancing solution using MEMS and BLE can be created. It is based on the RSSI measurements with different tilt and distance around a central node as in Figure 7.12. A key aspect of this solution is that it does not require availability of Internet for detecting the proximity between two nodes and alerts the users about another device being in close proximity. This has been especially found useful in COVID-19 situation where social distancing has become a norm. This scheme lends itself favorably to indoor navigation and this combination can be used for tracing of contacts and similarly any assets if so required (Singhal & Tiwari, 2020).

The Bluetooth protocol supports three topologies (Figure 7.13). First one is the pairing topology, where there is a one-to-one connection in which one device is called a master and the other is called a slave. One master can be connected to multiple slave devices thus giving rise to a star connection. Another useful topology is called broadcast topology where a device called broadcaster generates advertisement packets which can be received by the observers present in the same area. The broadcaster does not have any knowledge about the receivers. The third topology is the mesh topology where some nodes have the relay capability and can transmit the packets received from their neighbors. Thus, a packet can reach a farther in the network than the direct radio range. The mesh topology is most useful for the IoT networking. The biggest selling point of Bluetooth low energy as an IoT technology is that it's already supported by all modern smartphones, laptops, and tablets.

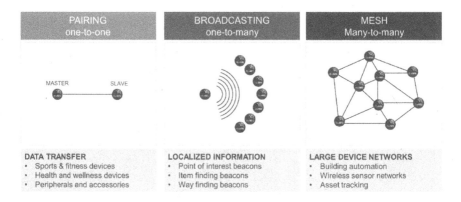

FIGURE 7.13 Bluetooth networking evolution. (From STMicroelectronics www.st.com.)

7.7.2.4 ZigBee

ZigBee is a low-cost, low power, standard targeted at the long battery life devices in wireless control and monitoring applications. ZigBee devices have low latency. ZigBee is a communication protocol formulated by the task force under the IEEE 802.15 working group operating in the ISM band. ZigBee builds on the physical layer and media access control defined in IEEE standard 802.15.4 for low-rate WPANs (TEC, 2017a). The typical uses of ZigBee are in the applications requiring low data rate, long battery life, secure networking, inexpensive, and self-organizing mesh network that can be used for industrial control, embedded sensing, medical data collection, smoke and intruder warning, building and home automation (including lighting), etc. (Varghese et al., 2019).

ZigBee supports both full function devices (FFD) and reduced function devices (RFD). RFDs are intended for very simple devices such as light switches. FFDs can perform communications routing and forwarding functions. The various roles in the network are:

- *ZigBee coordinator (ZC):* which initiates and configures network formation; acts as ZigBee router once the network is formed; is a full functional device.
- *ZigBee router (ZR):* participates in multi-hop routing of messages in mesh and cluster-tree networks; associates with ZC or with previously associated ZR in cluster-tree topologies; is a full functional device.
- *ZigBee end-device (ZED):* does not allow other devices to associate with it; does not participate in routing; is just a sensor/actuator node; can be an RFD.

A network shall include at least one FFD, operating as the PAN coordinator. The FFD can operate in three modes: a personal area network (PAN) coordinator, a router, or a device. A RFD is intended for simple applications that do not need to send large amounts of data. A FFD can talk to RFDs or FFDs while a RFD can only talk to a FFD (Kumar, 2010).

Communication Technologies for M2M and IoT Domain

The specification defines four key components (ZigBee Alliance, 2015):

1. Network layer
2. Application layer
3. ZigBee device objects (ZDOs)
4. Manufacturer defined application objects which allow for customization

ZDOs are responsible for tasks like keeping track of device roles, managing requests to join a network, as well as device discovery and security. The ZigBee network layer natively supports both star and tree networks, and generic mesh networking. Every network must have one coordinator device, tasked with its creation, the control of its parameters, and maintenance. Both trees and meshes allow the use of ZigBee routers to extend communication at the network level.

7.7.2.5 6LoWPAN

6LoWPAN is an open standard developed by IETF. It is an IPv6 adaptation layer running on top of IEEE 802.15.4 standard. A certain type of constrained networks called Low-Power Wireless Personal Area Networks (LoWPANs) has been widely used in a variety of applications in the field of IoT, including wearable or implantable devices, urban monitoring, control of large buildings, and industrial control applications. The LoWPANs are not limited to personal usage only. The most popular LoWPAN implementations are IEEE 802.15.4 defined by the IEEE 802.15 working group and Bluetooth Low Energy (BLE) developed by the Bluetooth Special Interest Group. Several network protocols based on IEEE 802.15.4 like ZigBee and 6LoWPAN have been developed and standardized. 6LoWPAN brings IPv6 to the constrained IoT networks, reusing the time-proven TCP/IP protocols on IoT, making it a natural and future proof choice for IoT network protocols (Figure 7.14) (Yang & Chang, 2019).

6LoWPAN stands for IPv6 over Low-Power Wireless Personal Area Networks and is an IP (Internet Protocol)-based technology. It is a network protocol that defines encapsulation and header compression mechanisms for the IP header. The standard has the freedom to choose frequency band and physical layer and can also be used across Ethernet, Wi-Fi, 802.15.4, and sub-1 GHz ISM. A key attribute is the

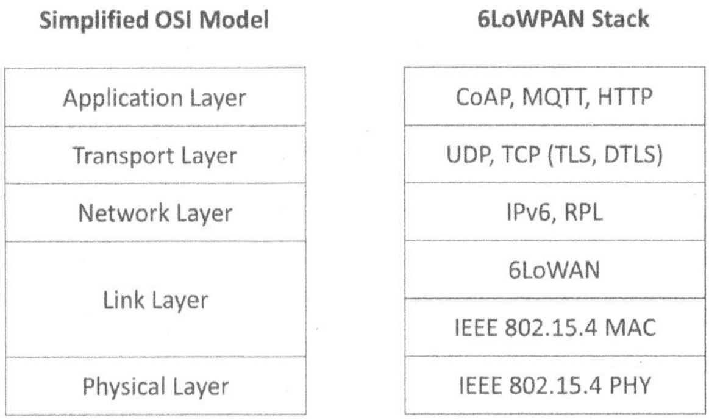

Simplified OSI Model	6LoWPAN Stack
Application Layer	CoAP, MQTT, HTTP
Transport Layer	UDP, TCP (TLS, DTLS)
Network Layer	IPv6, RPL
Link Layer	6LoWAN
	IEEE 802.15.4 MAC
Physical Layer	IEEE 802.15.4 PHY

FIGURE 7.14 6LoWPAN stack overview. (From Yang & Chang, 2019.)

IPv6 (Internet Protocol version 6) stack, which has been a very important introduction in recent years to enable the IoT. IPv6 is the successor to IPv4 (2^{32} addresses) and offers approximately 2^{128} addresses (approximately 3.4×10^{38}), enabling any embedded object or device to have its own unique IP address and connect to the Internet. Designed to send IPv6 packets over IEEE 802.15.4-based networks and implementing open IP standards including TCP, UDP, HTTP, COAP, MQTT, and web sockets, the standard offers end-to-end addressable nodes, allowing a router to connect the network to IP. 6LowPAN is a mesh network that is robust, scalable, and self-healing. Mesh router devices can route data destined for other devices, while nodes are able to sleep for long periods of time. Currently, Contiki, Thread, etc. are 6LoWPAN supported solutions (TEC, 2017a).

The salient characteristics of this technology are:

- Small packet size
- Low bandwidth (250/40/20 kbps)
- Topologies include star and mesh
- Low power, typically battery operated
- Easy translation to and from IPv6 packets
- Choice of physical layers

There are numerous open source and commercial implementations of 6LoWPAN. The most popular ones are: Contiki OS (Figure 7.15), TinyOS, Linux, and Openthread. They have been ported to a wide range of hardware platforms and architectures (Yang & Chang, 2019).

7.7.2.6 Z-Wave

Z-Wave is a wireless communications protocol used primarily for home automation. An IoT hub or gateway is a key component of a Z-Wave deployment. The hub

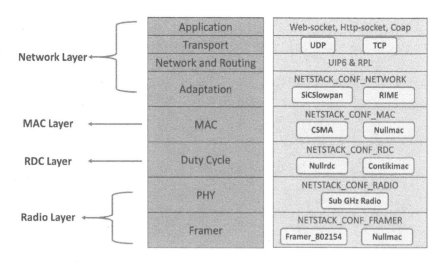

FIGURE 7.15 IP stack architecture in Contiki-based 6LoWPAN system. (From TEC, 2017a.)

communicates with the smartphone and/or the communication router to allow access to the home automation devices remotely. The network can support hundreds of sensors, devices, and actuators, which need to send data to a server and/or receive commands for configuration changes or action. The size of the property being managed by a Z-Wave hub can be quite large because Z-Wave uses "mesh network." Utility powered Z-Wave products act as relay nodes passing the messages along until the destination product is reached.

Each Z-Wave network has a unique ID, which is assigned to the Z-Wave hub and to every device in the network. This ensures that the neighbor's hub cannot control other's devices. When an extra level of security is needed, such as for door locks and other high security devices, Z-Wave has another level of security which uses AES128 encryption. Bluetooth mesh also has a similar mechanism called application keys (TEC, 2017a). Z-Wave uses a mesh network with the number of connected devices in a network limited to up to 232 nodes (Marksteiner et al., 2017).

7.7.2.7 EnOcean

EnOcean is one of the newest technologies in home automation, mainly aimed at zero energy consumption through energy harvesting. The unique beneficial feature of EnOcean devices is their ability to work battery-less and still having the ability to communicate wirelessly. This is achieved by means of micro energy converters along with ultra-low power electronics.

Early designs of EnOcean devices used piezo electric generators but were later replaced by electromagnetic energy sources. Because the devices are self-powered, the maintenance is minimal. Radio interference is also minimal as it operates in the less crowded 315 MHz band (Withanage et al., 2014).

7.7.2.8 ANT/ANT+

ANT represents another ultra-low-power, short-range wireless technology designed for sensor networks and similar applications. It, too, operates in the 2.4 GHz ISM band. This protocol is developed and sold by Canadian company Dyna stream Innovations Inc. It defines a wireless communications protocol stack that enables hardware operating in the 2.4 GHz ISM band to communicate by establishing standard rules for coexistence, data representation, authentication, and error detection. It is like Bluetooth low energy but is oriented toward usage with sensors. ANT uses the very short duty cycle technique and deep sleep modes to ensure very low power consumption. The ANT protocol is set up to use a single 1 MHz channel for multiple nodes due to a time division multiplex technique. Each node transmits in its own time slot. Modulation is GFSK. As it shares some aspects with Bluetooth Low Energy, it is comparatively easy to add ANT support to radios already supporting Bluetooth (TEC, 2017a). Some smartphones in the market natively support ANT protocol.

7.8 IoT HUBS AND GATEWAYS

Some of the protocols discussed in this section are not IP based, thus need to be bridged to IP networks using hub/gateway devices. Some hubs have multiple radios in them (Z-Wave, Bluetooth, etc.) so that they can perform different functions and support different products. Main functionalities provided by a gateway can be classified as follows:

  a. Protocol translation to connect variety of devices
  b. Perform data encryption for security
  c. Edge analytics
  d. Sensor supervision
  e. Provide cloud identity for the deployment
  f. Local data storage in case of connectivity is lost temporarily

An IoT Home gateway that can combine Bluetooth, sub-GHz, and Wi-Fi can be designed as an example of an IoT gateway (TEC, 2017b). IoT gateway helps connect things to broader Internet by using connectivity technologies suitable for resource limited things. There are a myriad of technologies and protocols available to communicate between things, gateway, and Cloud Application. A number of considerations are involved while designing a gateway:

- *Backend connectivity:* The IoT gateway may use short-range radio technology to connect to the IoT nodes but a long-distance link is needed to connect to the Internet. This selection is based on bandwidth requirements, available connectivity options in the area, and criticality of the application.
- *Management server:* IoT nodes are not generally accessed (through the gateway) on the Internet as standalone entities. It's more prevalent to have a central server managing the nodes, while IoT gateway facilitates this communication. We need to identify protocols for communication with the management server.
- *Local intelligence:* The IoT gateway can take most of the decisions locally and send only the filtered data to the cloud. This can make the system more efficient. The gateway decision logic may be programmed by the server for flexibility.
- *Power considerations:* The power source of gateway also affects our decisions related to the above points. As sensor networks become more prevalent and embedded in things, they would need to be as unobtrusive as possible and scavenge power from its environment.
- *Security:* This is a factor that can make or break the success of large-scale IoT networks. As these networks become part of application (some of them critical in nature), security will assume paramount importance.
- *Serviceability:* There must be a provision to service and to update the IoT gateway (and nodes) in the field. There should not be sole dependence of remote serviceability and we should have additional connectivity options to service the installation.

A detailed description of IoT Home gateway consisting of NFC, Wi-Fi, sub-Ghz Contiki, Bluetooth can be found in Annexure 2 of TEC (2017b). The gateway can be controlled using an APP and Bluetooth locally or connected to cloud using Wi-Fi.

7.9 TOPOLOGIES: MESH VS. STAR

Mesh networks have some benefits over star networks; for example, they are able to cover greater distances even if the underlying communication technology has limited range. Also, mesh technologies are more robust than star because there is no single point of failure. The penalty of using a mesh network over the star network comes in the form of network congestion in the network and higher latency. Mesh networks consume more energy because the relay nodes must be active all the time listening to the transmission from other nodes and retransmitting it.

7.10 IoT COMMUNICATION PROTOCOLS

Communication protocols are of great importance in heterogeneous systems, which define an accepted framework for device interaction. Therefore, some application layer protocols have been designed for M2M communication and IoT systems, particularly, Message Queuing Telemetry Transport (MQTT), Advanced Message Queuing Protocol (AMQP), and Constrained Application Protocol (CoAP) (Moraes et al., 2019).

There are some protocols for IoT application layer, which focus on issues related to M2M communication.

- *Message Queue Telemetry Transport* (MQTT) is a message protocol designed to work on devices that have constrained computational resources. It is one of the lightweight messaging protocols that follows the publish-subscribe paradigm, which makes it rather suitable for resource-constrained devices and for non-ideal network connectivity conditions, such as with low bandwidth and high latency. Because of its simplicity, and a very small message header comparing with other messaging protocols, it is often recommended as the communication solution of choice in IoT. Since it was designed to be as lightweight, MQTT does not provide encryption, and instead, data is exchanged as plain-text, which is clearly an issue from the security standpoint. Therefore, encryption needs to be implemented as a separate feature, for instance, via TLS, which on the other hand increases overhead (Dizdarević et al., 2019). The protocol adopts a flexible routing mechanism and asynchronous communication based on the publish-subscribe paradigm. MQTT works over TCP and it has three quality of service (QoS) levels:
  - QoS-0, the message is sent only one time without confirmation.
  - QoS-1, the message sent is guaranteed to be delivered (maybe with duplicates).
  - QoS-2, only one message is assured to be delivered to subscribers without duplicates.

- *Advanced Message Queuing Protocol* (AMQP) is an open application layer protocol for IoT. AMQP supports publish-subscribe and client-server communication and it has similar QoS mechanisms as MQTT. The protocol also operates using TCP. AMQP uses TCP for reliable transport, and in addition it provides three different levels of QoS, same as MQTT. The AMQP provides complementary security mechanisms, for data protection by using TLS protocol for encryption, and for authentication by using SASL (Simple Authentication and Security Layer). This protocol is better suited in the parts of the system that is not bandwidth and latency restricted, with more processing power (Dizdarević et al., 2019).
- *Constrained Application Protocol* (CoAP) was mainly developed to interoperate with Hypertext Transfer Protocol (HTTP). Hence, CoAP has some functionalities resembling HTTP, such as GET and POST, but it utilizes UDP. As UDP is inherently not reliable, CoAP provides its own reliability mechanism, adopting Confirmable (CON) and Non-Confirmable (NON) messages. A Non-Confirmable message does not require a confirmation about its delivery. When a server is not able to process a Non-Confirmable message, it may reply with a reset message (RST). If an acknowledgment is required, a Confirmable message is adopted. Besides, CoAP may provide two types of response (Figure 7.16):
  - Piggybacked and Separate response.
  - In piggybacked, the response returns in the acknowledgement (ACK) packet of the request. There is no need for separately acknowledging a piggybacked response, as the client will retransmit the request if the ACK packet carrying the piggybacked data is lost.
  - In separate response, the acknowledgement and response messages are sent in different packets. More specifically, when a server processes the request, it sends a response packet. The client then transmits an acknowledgement message due to the receiving of the server response.

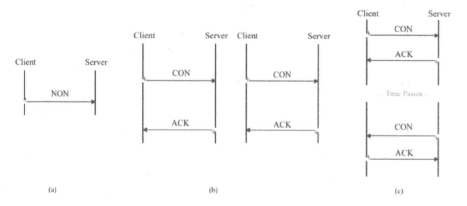

FIGURE 7.16 CoAP (a) non-conformable, (b) piggybacked, and (c) separate response. (From Moraes et al., 2019).

Separate response is ideal for situations in which the server cannot immediately respond to a request in a confirmation message. The server will respond with an ordinary acknowledgment.

7.11 CONCLUSION

The IoT has the potential to revolutionize and change the way all businesses, governments, and consumers interact with the physical world. This level of disruption will have a significant impact on the world by improving the quality of life. Pervasive nature of IoT devices presents some unique challenges related to privacy protection of its users. Another challenge is potential interoperability issues due to the inherent heterogeneous nature of the IoT devices. However, full potential of IoT can be realized if industry and governments around the world join hands to come up with light-handed regulations and nudge the manufacturers to create interoperable and secure IoT devices.

To conclude this chapter, it is important to mention that IoT communication technologies and installed nodes are increasing at rapid rate. Some of the key communication technologies were briefly described in this chapter, but many others had to be left out in the interest of the length and also their wider familiarity. Wi-Fi and DSL and host of other technologies provide very useful backbone for the IoT/M2M devices on the gateway side.

The trend in IoT with ever increasing number of nodes is on security and this dimension plays counterintuitively for the computing, battery life, and simplicity of the edge nodes. The M2M/IoT devices, where more security is needed, would see more and more usage of learning at the edge and blockchain technologies for agility and security as well.

REFERENCES

Ahuja, S., & Potti, P. (2010). *An Introduction to RFID Technology. 2010* (August), 183–186. https://doi.org/10.4236/cn.2010.23026

Ayoub, W., Samhat, A. E., Nouvel, F., Mroue, M., & Prévotet, J. C. (2019). Internet of mobile things: Overview of LoRaWAN, DASH7, and NB-IoT in LPWANs standards and supported mobility. *IEEE Communications Surveys and Tutorials*, *21*(2), 1561–1581. https://doi.org/10.1109/COMST.2018.2877382

Coskun, V., Ozdenizci, B., & Ok, K. (2013). A survey on near field communication (NFC) technology. *Wireless Personal Communications*, *71*(3), 2259–2294. https://doi.org/10.1007/s11277-012-0935-5

Dizdarević, J., Carpio, F., Jukan, A., & Masip-Bruin, X. (2019). A survey of communication protocols for internet of things and related challenges of fog and cloud computing integration. *ACM Computing Surveys*, *51*(6), 1–30. https://doi.org/10.1145/3292674

Gomez, C., Oller, J., & Paradells, J. (2012). Overview and evaluation of Bluetooth low energy: An emerging low-power wireless technology. *Sensors (Switzerland)*, *12*(9), 11734–11753. https://doi.org/10.3390/s120911734

ITU-T. (2012). An overview of internet of things. In *Overview of the Internet of Things*.

Kumar, M. (2010). ZigBee: The low data rate wireless technology for ad-hoc and sensor networks. *NCCI 2010 - National Conference on Computational Instrumentation CSIO Chandigarh*, India, 19–20 March 2010.

Landt, J. (2005). The history of RFID. *IEEE Potentials, 24*(4), 8–11. https://doi.org/10.1109/MP.2005.1549751

Lavric, A., Petrariu, A. I., & Popa, V. (2019). Long range SigFox communication protocol scalability analysis under large-scale, high-density conditions. *IEEE Access, 7*, 35816–35825. https://doi.org/10.1109/ACCESS.2019.2903157

LoRa® Alliance. (2015). What is it ? A technical overview of LoRa. *LoRa Alliance, November.* https://lora-alliance.org/sites/default/files/2018-04/what-is-lorawan.pdf

Marksteiner, S., Jimenez, V. J. E., Valiant, H., & Zeiner, H. (2017). An overview of wireless IoT protocol security in the smart home domain. *Joint 13th CTTE and 10th CMI Conference on Internet of Things – Business Models, Users, and Networks, 2018-January*, 1–8. https://doi.org/10.1109/CTTE.2017.8260940

Mehmood, Y., Görg, C., Muehleisen, M., & Timm-Giel, A. (2015). Mobile M2M communication architectures, upcoming challenges, applications, and future directions. *EURASIP Journal on Wireless Communications and Networking.* https://doi.org/10.1186/s13638-015-0479-y

Mekki, K., Bajic, E., Chaxel, F., & Meyer, F. (2018). Overview of cellular LPWAN technologies for IoT deployment: Sigfox, LoRaWAN, and NB-IoT. *2018 IEEE International Conference on Pervasive Computing and Communications Workshops, PerCom Workshops 2018*, 197–202. https://doi.org/10.1109/PERCOMW.2018.8480255

Moraes, T., Nogueira, B., Lira, V., & Tavares, E. (2019). Performance comparison of IoT communication protocols. *Conference Proceedings – IEEE International Conference on Systems, Man and Cybernetics, 2019* (October), 3249–3254. https://doi.org/10.1109/SMC.2019.8914552

Peña Queralta, J., Gia, T. N., Zou, Z., Tenhunen, H., & Westerlund, T. (2019). Comparative study of LPWAN technologies on unlicensed bands for M2M communication in the IoT: Beyond Lora and Lorawan. *Procedia Computer Science, 155*(2018), 343–350. https://doi.org/10.1016/j.procs.2019.08.049

Qin, Z., Li, F. Y., Li, G. Y., McCann, J. A., & Ni, Q. (2019). Low-power wide-area networks for sustainable IoT. *IEEE Wireless Communications, 26*(3), 140–145. https://doi.org/10.1109/MWC.2018.1800264

Satan, A. (2018). Bluetooth-based indoor navigation mobile system. *Proceedings of the 2018 19th International Carpathian Control Conference, ICCC 2018*, June, 332–337. https://doi.org/10.1109/CarpathianCC.2018.8399651

Singhal, I., & Tiwari, R. (2020). Introduction social distancing detection using Bluetooth ® low energy. *STMicroelectronics, AN5508, 0*(June 2020), 1–17. www.st.com

Sinha, R. S., Wei, Y., & Hwang, S. H. (2017). A survey on LPWA technology: LoRa and NB-IoT. *ICT Express, 3*(1), 14–21. https://doi.org/10.1016/j.icte.2017.03.004

Tabbane, S. (2018). *IoT Standards. Part II: 3GPP Standards. September,* 130. https://www.itu.int/en/ITU-D/Regional-Presence/AsiaPacific/Documents/Events/2018/IoT-BDG/7. IoT Standards Part II - Sami Tabbane.pdf

TEC. (2017a). *Technical Report: Communication Technologies in M2M/IoT Domain, TEC-TR-IoT-M2M-008-01.* https://tec.gov.in/pdf/M2M/Communication Technologies in IoT domain.pdf

TEC. (2017b). Technical report M2M/IoT enablement in smart homes TEC-TR-IoT-M2M-007-01 M2M smart homes working group. In *Smart Homes* (Issue March). https://doi.org/10.5771/9783845257723

Varghese, S. G., Kurian, C. P., George, V. I., John, A., Nayak, V., & Upadhyay, A. (2019). Comparative study of ZigBee topologies for IoT-based lighting automation. *IET Wireless Sensor Systems, 9*(4), 201–207. https://doi.org/10.1049/iet-wss.2018.5065

Withanage, C., Ashok, R., Yuen, C., & Otto, K. (2014). A comparison of the popular home automation technologies. *2014 IEEE Innovative Smart Grid Technologies - Asia, ISGT ASIA 2014*, 600–605. https://doi.org/10.1109/ISGT-Asia.2014.6873860

Yang, Z., & Chang, C. H. (2019). 6LoWPAN overview and implementations. *Proceedings of the 2019 International Conference on Embedded Wireless Systems and Networks*, 357–361. https://dl.acm.org/doi/10.5555/3324320.3324409

Yousuf, A. M., Rochester, E. M., Ousat, B., & Ghaderi, M. (2019). Throughput, coverage and scalability of LoRa LPWAN for Internet of things. *2018 IEEE/ACM 26th International Symposium on Quality of Service, IWQoS 2018*. https://doi.org/10.1109/IWQoS.2018.8624157

ZigBee Alliance. (2015). ZigBee specification. In *ZigBee Alliance*. ZigBee Alliance. http://ieeexplore.ieee.org/document/6264290/

8 Security Challenges and Solutions in IoT Networks for the Smart Cities

A. Procopiou and T.M. Chen
University of London
London, United Kingdom

CONTENTS

8.1 Introduction 161
8.2 IBM Smart City Architecture 162
8.3 IoT Integration in Smart Cities 164
8.4 IoT in Smart Cities Technologies 169
    8.4.1 IoT Wireless Technologies 169
    8.4.2 IoT Wireless Technology Standards and Protocols 170
8.5 IoT In Smart Cities Security Analysis 170
    8.5.1 IoT in Smart Cities Security Vulnerabilities 170
    8.5.2 IoT in Smart Cities Threats and Attacks 173
        8.5.2.1 IoT Hardware Attacks 173
        8.5.2.2 IoT Attacks on the Network 174
        8.5.2.3 IoT DDoS Attacks 175
        8.5.2.4 IoT Attacks on Authentication 177
        8.5.2.5 IoT Malicious Software 178
8.6 IoT Attacks Countermeasures 179
    8.6.1 Countermeasures on IoT Hardware Attacks 179
    8.6.2 Countermeasures on IoT Attacks on the Network and DDoS Attacks.. 184
    8.6.3 Countermeasures on IoT Authentication and Illegal Modification Attacks 186
    8.6.4 Countermeasures Against IoT Malicious Software 189
8.7 Discussion and Future Directions 192
8.8 Conclusion 196
References 197

8.1 INTRODUCTION

According to the United Nation's revision of world urbanization prospects report, it is estimated that by 2050, more than 68% of the population is going to live in cities [1] with approximately 2.5 billion people and 2.9 billion vehicles. Hence, owing to global overpopulation, there is naturally an increase in the demand for water, electricity, and gas. Furthermore, a greater demand for transportation systems, the health sector, and safety services is evident. Specifically, according to Mohanty et al. [2], currently, cities consume 75% of the world's resources and energy, which leads to the generation of 80% of greenhouse gases.

Furthermore, this urban overpopulation causes numerous other problems including environmental pollution, emissions exhaustion, global warming, energy waste, health problems, public transport congestion, and overcrowding. Therefore, the transformation of regular cities into smart cities is more vital than ever.

The main objectives of a smart city are to improve the lives of its citizens, promote sustainability, and protect the environment [3]. It aims to achieve that through the usage of intelligent technologies such as the cloud, blockchain, data analytics, artificial intelligence, machine learning, sensors/actuators, and the Internet of Things (IoT) to optimize the functioning of its core components and make them "smarter." Examples of these components are transportation systems, the health system, energy and waste infrastructures, residential premises, industrial, commercial, and governmental sectors.

A wide set of complex systems and infrastructures, technologies, social and political mechanisms, the economy, and the citizens are going to play a vital role in its realization, with the most popular being the IoT.

IoT forms the interconnection of physical entities with electronics, software, sensors, and actuators that are constantly connected to the Internet and collect and exchange data either with other intelligent devices located in the same network or other networks. Through these processes, they offer a set of cyber and physical activities. Also, IoT offers the ability for things to be monitored and/or controlled remotely, which increases efficiency and accuracy, reduces costs and change of errors, and increases human safety as in the case of an accident, no human lives are going to be directly impacted or put at risk. Using IoT in smart cities allows for the vast collection of data from different and multiple types of sensors from anyone at any time and from anywhere, which is transferred to the cloud to be processed further and to monitor the state of various networks and infrastructures of the smart city, consequently assisting its citizens with their lives.

However, the presence of IoT in smart cities brings a whole lot of vulnerabilities that, when exploited by adversaries, can result in not only data theft and network malfunction, but disruptions to the life of its citizens which can be hazardous to them and even cost human lives. In more detail, the constant connection to the Internet, the continuous interaction among different assets, the heterogeneity in terms of protocols and functioning across different IoT devices, IoT's weak security measures, and the interconnection of the different networks together make the possibility of a cyberattack more than likely.

In this chapter, we aim to explore the security and privacy issues emerging from the deployment of IoT in smart cities and its various networks and infrastructures, and what solutions could be provided to guard against them.

8.2 IBM SMART CITY ARCHITECTURE

Owing to the smart city's complex and heterogeneous nature both at technical as well as conceptual levels, proper categorization and accurate modularization of its different components are necessary. Many frameworks have been proposed both at

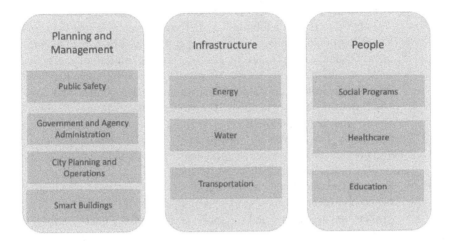

FIGURE 8.1 IBM smart city architecture.

conceptual as well as technical levels. In this chapter, IBM's Smart City Conceptual Framework is presented and described [4] since we have concluded that it is an ideal representation of the core components a smart city should encompass, as illustrated in Figure 8.1. The framework is divided into three dimensions: Planning and Management, Infrastructure, and People. Each dimension has a number of subcategories, which are defined and explained below.

1. Planning and Management: Through this dimension, it is ensured that citizens are aware of the activities they can perform while still performing their frequent usual duties.
    a. Public Safety: Public Safety Services can collect data regarding traffic, weather, crime, health issues, natural disasters, hazardous materials, and so on, and analyze them. This analysis can be sent to relevant stakeholders, such as law agencies and governments, for further analysis. Examples of such smart systems are smart lighting systems, surveillance systems, weather monitoring systems, fire and detection systems, earthquake detection and flood prevention systems, air pollution monitoring systems as well as health monitoring systems for people in need.
    b. Government and Agency Administration: Government policies are to be reshaped so effective smart city growth is possible that will maintain the citizens' needs. Examples of such systems are e-voting systems and online tax-paying systems.
    c. City Planning and Operations: Through the effective deployment of smart systems and data analytics, a smart city can effectively manage its daily operations. Some examples are urban noise systems, squander management systems, smart parking systems, and traffic overcrowding systems.

d. Smart Buildings: Through the concept of IoT, smart technologies are introduced to traditional buildings. Through sustainable materials and smart sensors, a number of different operations can be performed efficiently. Some examples are measuring, sensing, and assessing buildings' condition, energy usage, demand/response, lighting, fire/flood incidents, elevators, air quality, security and access control, and computer networks and applications. The most popular examples are Smart Homes and Smart Buildings.
2. Infrastructure: This dimension aims to make the city stand on its own in order to provide useful services to its citizens. However, a livable smart city has its own necessities that have to be fulfilled.
  a. Energy: The traditional electrical grid, providing electricity from utilities to consumers' homes, is to be replaced with the smart grid. Smart grid encompasses advanced communication networks and big data analytics. Through them, utilities can manage demand and response and enable better and more efficient energy consumption through intelligent distribution management systems. It also enables two-way communication between consumers' home and utilities so maximum efficiency can be achieved and more effective feedback can be exchanged between the two ends.
  b. Water: Through smart water waste management systems better efficiency of water management can be achieved. The systems can visualize consumption data, water and quality of flow, and pressure.
  c. Transportation: Through the deployment of sensors such as location detection systems and cloud-based real-time analytics in all types of transportation infrastructures, traffic congestion and transport times can be reduced.
3. People: This dimension is about the citizens of the smart city and how their interactions and own services provided to other citizens can be enhanced and expanded through the services a smart city offers.
  a. Social Programs: Solutions deployed in this domain will give a chance to citizens to take part in social programs so they can have better quality of lives.
  b. Healthcare: Healthcare services need to be fast, interconnected, efficient and accurate, and intelligent so that accurate diagnoses can be made and medical staff can achieve their goals (cure diseases and illnesses).
  c. Education: Through smart education services, effective teaching can be achieved and can be expanded outside classrooms.

8.3 IoT INTEGRATION IN SMART CITIES

For the automatic capture and sharing of smart city data in real time, interaction of smart city systems with its citizens, and the communication of such systems with dedicated control servers and the cloud in the smart city, one of the most vital

Security Challenges and Solutions in IoT Networks for the Smart Cities

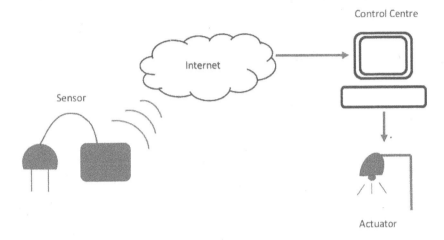

FIGURE 8.2 IoT technologies components.

components is the adoption of the IoT. IoT includes the set of technologies and machinery that enable everyday objects to become "smart."

The definition of an object becoming "smart" includes its ability to collect data from its surrounding environment in real time and share this data and information (mostly wirelessly) with other "smart" objects and dedicated servers/cloud, being connected to the Internet. For the IoT to be successfully integrated and enabled, multiple technologies are used including sensors and actuators, embedded systems, machine learning, and data analytics. A visual representation of how these technologies interact with each other is illustrated in Figure 8.2.

Sensors are installed in a smart city network environment for the effective and accurate detection of event changes in the environment as well as the collection of relevant data that could assist in smarter decision-making. Sensors can operate either with an external power source or on their own. Sensors can be a part of an embedded system that includes a type of processor that performs basic validation and processing before forwarding it to the relevant servers.

Through the usage of a (most of the time) wireless medium, this data is forwarded to a dedicated server, which validates the input and uses this data accordingly for smart decision-making, predictions, or anomaly detection. The output assists in the potential change of the environment/system/network state through specific actions with the help of relevant algorithms, data analytics, and machine learning. In practice, these changes are made through the use of actuators. An actuator is a component that controls a machine or system upon receiving a signal, so a task or an operation is performed.

To summarize, the usage of IoT successfully unites the physical with the digital world, connecting, integrating, and utilizing data from one device to the other, and takes beneficial decisions for the city, the environment, and, most importantly, the citizens. Further, we present and describe in practice how IoT assists in the functioning of smart city services and what benefits are gained through them [5–8].

1. Health Building and Bridges Monitoring: Buildings (especially historical ones) and bridges require continuous and in-depth monitoring to maintain their conditions and interfere if any damage is likely to occur due to external agents. Through the integration of IoT, dedicated sensors can be deployed to collect relevant data regarding vibration, deformation, building stress, atmospheric state of nearby areas, pollution levels, temperature, and humidity so that a detailed state of the environment can be known [9]. As a result, multiple benefits are gained from this deployment including the proactive maintenance and immediate restoration of damaged parts, better understanding of the impact on buildings from natural disasters (e.g., earthquakes), and sharing of this data with relevant assets.
2. Energy and Water Consumption Management: Using IoT in energy systems such as the Smart Electrical and Water Grids, we can monitor the energy/water consumption in a smart city and become more aware of the actual needs of citizens individually or based on the geographical area. Hence, effective optimization and prioritization of these resources can be achieved so that energy loss is minimized and damage to equipment is avoided. Furthermore, with the usage of IoT devices in such critical infrastructures, the effective and reliable use of such important cyber-physical systems is more direct and effective, helping in immediate detection and repair of any malfunction. Finally, the introduction of alternative energy generation means (e.g., photovoltaic panels) becomes easier and the possible exchange of energy between utilities and consumers is potentially more direct and straightforward [10, 11].
3. Waste Management: Unarguably, waste management forms a big issue in modern cities of today. Therefore, the deployment of IoT, such as smart waste containers, can optimize the waste collection, reduce cost, and improve recycling [12]. This could potentially help in significant economical savings, substantial protection of the environment, and improvement of the citizens' lives and well-being.
4. Environmental Monitoring: IoT can be deployed directly on the environment for effective monitoring of various environmental assets. One important example is air quality, as its effective monitoring can help individuals with or without pre-existing lung conditions to optimize their outdoor activities. IoT networks can be developed to detect CO_2 emissions from vehicles, factories, and other gases and sources. Another example includes forest fire detection where IoT would be able to immediately detect fire risk conditions before they spread to forests and most importantly residential sites and notify the relevant services quickly and accurately. In addition, other natural disasters such as floods and earthquakes can be detected through the usage of IoT vibration and earth density sensors and notify the relevant services for immediate action.
5. Noise Monitoring: Excessive noise levels are classified as a form of acoustic pollution and have been an actual problem in various cities around the world such as Barcelona. IoT devices can be deployed to effectively monitor these levels. Furthermore, noise monitoring systems could potentially be

used for enhancing public security and, therefore, increase people's sense of physical security [13].
6. Smart Lighting: IoT can be used to optimize street lighting systems in various ways. Examples include optimizing street lamps' intensity based on day/time, season, current weather conditions as well as presence and load of vehicles and/or pedestrians, which results in energy saving. Furthermore, both drivers and pedestrians can be assisted in their journeys regardless of the weather conditions. Finally, smart street lights can be potentially used as Wi-Fi access points to provide Internet connection to citizens.
7. Smart Traffic Lights and Traffic Congestion: Due to overpopulation in large cities, traffic congestion is considered a big issue. Through the deployment of various IoT sensors such as vehicle detection, air quality, and acoustic sensors on roads as well as GPS installed in smart vehicles, real-time status of the traffic can be monitored. Through this and the effective usage of traffic lights, a considerable set of issues can be solved such as minimization of traffic congestion during rush hours/busy roads, recommendation of alternative routes to drivers as well as immediate action in case of a road accident.
8. Smart Parking Areas: Through the usage of IoT parking sensors and intelligent displays, finding a parking slot will become easier and less frustrating [14, 15]. There benefits from the deployment of IoT are multiple including the minimization of CO_2 emissions, lesser traffic congestion, less time spent in locating a parking slot as well as the citizens being overall happier. In addition, through the usage of short-range wireless communication technologies such as Radio Frequency Identification (RFID) and Near Field Communication (NFC), the effective monitoring of parking permits (e.g., for residents and/or disabled people) could be achieved, thus providing a more fair and a better quality of life for every smart citizen.
9. Public Health Services: IoT can revolutionize the public health industry as its deployment has multiple benefits to both the citizens and the medical staff involved. Firstly, remote healthcare will be provided to citizens who live in rural areas, have difficulty in coming to health centers or hospitals or their medical state does not allow them to move from their homes. Furthermore, even for people who can visit the health services provided by a smart city, the usage of mobile client services and apps will be beneficial both for them and the healthcare system. With the integration of IoT technologies, the citizens will now be quickly assessed and correct decisions will be made for them and the health system's risk of being overwhelmed will be minimized. In addition, citizens with chronic illnesses and conditions will be monitored 24/7 through the usage of IoT mobile medical devices (wearable, mobile, or implanted), so quick and reliable decisions can be made for them based on an extensive amount of real-time data collected regarding their conditions. Examples of such devices include blood pressure, heart rate, and glucose monitors, ECG and other remote physiological monitors, threshold-triggered alarm generators, and so on [16]. On the other hand, hospitals and medical centers will be transformed into smart ecosystems, through the usage of IoT

technologies. Smart medical equipment will improve diagnostics as they will have the capability of using patient data to choose the best treatment possible. Furthermore, both patients and medical staff can be tracked through various identification systems such as tags, bracelets, labels, and smart badges (e.g., ultrasound-enabled) so that activities can be optimized effectively [17].

10. Intelligent Transportation: In modern cities, private and public transportation systems are vital for the efficient operation of a city and its services. Transportation services can be public (railways, buses, metro, trolley buses, ferries, etc.) or private (airports, bike hires, taxis, cars, logistics/freights, etc.). They are also connected to non-transport operators and non-operators (e.g., emergency services, public safety, energy services, banks, and regulators) for assistance in their effective management and maintenance. Using IoT, safety, comfort, reliability, and optimization of services can be ensured in multiple ways. Firstly, through the deployment of smart sensors, relevant data regarding a commercial vehicle is collected which is essential for its proper care and maintenance; also useful data regarding the road conditions is collected which is equally important for the smart citizens' safety. Furthermore, based on its geographical location, passengers have a real-time status of the time it will take to reach their destination. In addition, effective traffic coordination and maintenance of lights can be achieved to avoid traffic congestion [18].

11. Public Safety: Beyond the integration of cyber-physical systems and the usage of IoT technologies for the transformation of regular services to intelligent versions of themselves, smart cities also provide physical safety and security to the citizens. IoT technologies can alert citizens in case of dangerous situations of emergency such as public shootings, enhance public safety through numerous technologies such as CCTV and the presence of monitor sensors, and assist in the protection of certain population groups such as women and children [19–25].

12. Assisted living and well-being in smart buildings: One of the most important aspects of the usage of IoT technologies in smart cities is turning all types of different buildings into intelligent buildings assisting the citizens. Examples include from simple residential homes and apartments to public buildings such as schools, universities, government administration offices, museums, and libraries and from medium-sized restaurants and retail stores to large stadiums, cinemas, theaters, and shopping malls. Through the deployment of IoT, the constant and effective monitoring of the environment can be achieved (e.g., temperature, humidity, air quality, human presence, and so on) and effective decisions regarding the functionality of the building can be made (e.g., heating, AC, lighting, and so on). Through this, an accurate representation of each building's state with regards to its occupants will be provided to the relevant people in charge and effective reduction of energy and resources can be achieved. Furthermore, assistance in navigating and using each building's facilities can be provided to individuals [15].

8.4 IoT IN SMART CITIES TECHNOLOGIES

Unarguably, the functioning of IoT networks presents significant differences in the functionality of various devices, resources, and their overall capabilities when compared to traditional information and communication technology (ICT) networks. Therefore, new technologies need to be designed and developed for the correct accommodation of this new type of network to their best interest. In this section, we present and discuss the most relevant technologies and protocols used in IoT based on the TCP/IP layer they belong to. Note that some of them were already used for other types of networks, while some have been developed specifically for the IoT.

8.4.1 IoT Wireless Technologies

RFID: RFID is one of the most popular technologies used in our everyday lives, for example, passports and credit cards. Using electromagnetic fields, RFID can identify and keep track of objects which have an RFID tag attached to them. An RFID tag can both receive and send data. When the RFID tag is within the reach of an electromagnetic signal sent by an RFID reader device, the tag transmits the data back to it. RFID is successfully used in IoT for various networks of the smart city such as smart agriculture, smart airports, and smart hospitals. Examples in smart agriculture include RFID tracking the potted plants in greenhouses and food traceability in the supply chain. In smart airports, RFID technology is used to track traveling luggage, and in smart hospitals to similarly locate assets (e.g., medical equipment), patients, and staff, optimize their schedule inventory management, and remote monitoring of the elderly [17, 26–29].

Wireless Sensor Networks (WSN): WSN is a large group of sensors used for monitoring an area's state and its environmental conditions, and forwards the data to a dedicated base station. WSNs can cover a large geographical area because of their distributive nature and autonomy. WSNs consist of "nodes" which are connected to one or more sensors. A node has a radio sender and receiver component, an antenna, a microcontroller, an electronic circuit for communicating with sensors, and an energy source, most of the time a battery. Their development was initially commenced by military applications and specifically battlefield surveillance [30]. Examples of WSN in IoT include the monitoring of critical infrastructures such as the smart grid or any other industrial cyber-physical systems [31–33].

NFC: NFC is a set of communication protocols for devices that are in a very close proximity to each other, typically 4 cm or less and they require low power and data rate requirement. In IoT, NFC-enabled devices can be identity documents or keycards with which individuals can be authenticated and authorized on activities [34]. Furthermore, NFC can be used in contactless payment systems such as mobile payment, thus replacing credit cards and electronic ticket cards. In addition, NFC can enable the sharing of files small in size such as contacts, and bootstrapping fast connections so that larger files can be shared such as photos and videos.

8.4.2 IoT Wireless Technology Standards and Protocols

Numerous wireless technology standards have been proposed for the effective and fast communication of resource-constrained devices and networks including IoT. Below, we briefly present and describe the most popular and widely used wireless technologies. We also have to point out other existing wireless standards used such as Wi-Fi, cellular (4G, LTE), 5G, and WiMax. However, in the context of this chapter, we have chosen to focus on wireless standards that are specifically tailored for resource-constrained devices [33].

> Bluetooth: Bluetooth is a wireless technology standard proposed and invented by Ericcson back in 1994. It aimed to exchange data over a short distance by using short-wavelength ultrahigh frequency (UHF) radio waves ranging from 2.4 to 2.485 GHz. The devices participating in this exchange can be either static or mobile [35].
> LoRa (Long Range): LoRa (also LoRaWAN) belongs to the family of low-power wide-area network protocols and has been developed by Cycleo of Grenoble (France) and acquired by Semtech. LoRa makes usage of different license-free sub-gigahertz radio frequency bands depending on the geographical region to achieve long-range transmissions, but by maintaining low power consumption; an ideal combination for IoT networks found in large geographical areas such as smart agriculture and sea-life monitoring [36].
> Zigbee: Zigbee is a high-level communication set of protocols used for constructing personal area networks that have small and low-power digital radios, firstly proposed in 1998 and then standardized in 2003. Popular examples of Zigbee usage include smart home automation, collection of data from medical devices, and any other small-scale applications of which wireless communication is a vital part and does not cover a large geographical area. Similarly to LoRa, Zigbee is a low power consumption protocol and its transmission distance is limited to only 10–100 m. However, Zigbee's transmission range can be enhanced when used in combination with a mesh network where additional devices running Zigbee can be used as beacons [35].

8.5 IoT IN SMART CITIES SECURITY ANALYSIS

8.5.1 IoT in Smart Cities Security Vulnerabilities

As IoTs are deployed in all the domains of a smart city, they bring unique characteristics that shape the concept of this new environment of multiple, complex, and heterogeneous networks. These characteristics bring new security challenges that must be realized. Furthermore, as cyber-physical systems are a vital part of a smart city, additional issues are introduced that have to do with the direct involvement of humans. The most important security vulnerabilities of IoT are described below, inspired and extended by Arasteh et al. [37–39].

1. Resource constraints: As known, most IoT devices demonstrate resource-constrained issues in multiple assets such as memory, CPU power, storage, and battery. Hence, they can be weak when it comes to storage and computational capabilities. Therefore, any already existing cryptographic, authentication, and intrusion detection countermeasures are more than likely unable to be deployed.
2. Energy constraints: IoT devices can operate either by electricity or battery. For the devices that are dependent on a battery, it is a challenge to deploy additional software and/or application as it is likely to drain the battery sooner and cause issues in the communication with other devices, storage of data, and calculations.
3. Heterogeneous communication protocols: Each IoT device comes with its own protocols and standards based on its concept, usage and usability, and functionality. Thus, in smart city networks and cyber-physical systems, the heterogeneity can vary based on the system, its functioning, and its priorities. This heterogeneity can potentially cause problems in the effective communication among different devices both inside and outside of their primary network. Furthermore, traffic analysis for potential threats can be proved more complex due to the heterogeneous traffic generated.
4. IoT device standards: IoT technology can still be considered new as it is continuously being developed and evolved. As a result, the technologies and concepts are still in the process of being standardized globally. This could potentially cause some issues in their reliability, functionality, and usage.
5. Cross-network interaction and interconnection: For the effective management of the smart city network and its services and the beneficial assistance to its citizens, various networks and cyber-physical systems deployed must have a constant and minimally interrupted interconnection and communication. However, such a tight interconnection among different systems can potentially raise issues such as cross-network cyberattacks, especially in cases in which one of the two networks involved is a smart infrastructure such as the smart grid and the other is a private network, such as smart neighborhood consisting of multiple smart homes. An adversary aiming to disrupt a smart city is likely to compromise private networks, where the security of the network is based on the owner's efforts hoping there are security loopholes, so they can reach a critical infrastructure with more ease compared to directly attacking a critical infrastructure.
6. Unreliable communications: Due to resource-constrained issues in IoT devices, communication protocols are chosen based on speed and lightweight nature rather than reliability. Such an example is the usage of user datagram protocol (UDP) in the constrained application protocol (CoAP) protocol, one of the most popular application layer IoT protocols used. As a result, potential issues regarding the successful delivery of packets can be observed and retransmission can be costly for IoT devices due to their resource constraints.
7. Physical access: IoT devices in a smart city are likely to be installed in large geographical areas, especially in smart agriculture and environmental

monitoring systems. It is highly unlikely for all of these devices to be periodically monitored and checked by human personnel. Therefore, there is a high chance for such devices to be physically tampered and/or damaged by adversaries.

8. Wireless machine-to-machine (M2M) communications [40]: IoT devices can directly exchange data with each other over a wireless medium in various scenarios such as air quality monitoring systems. Therefore, wireless communication combined with potentially weak security countermeasures can be an issue with regard to the confidentiality of data and authentication and trust of devices.
9. Nodes mobility: Mobility of devices in IoT networks is an important aspect in providing assisted living and essential help as well as monitoring surrounding environments, especially where human access is classified as hazardous for them. However, an increased mobility causes issues in the effective communication among devices and their base stations. Therefore, the fast and accurate detection of compromised devices can be a challenge in such IoT networks in a smart city.
10. Open APIs: Most technologies integrated into IoT devices are open and accessible by everyone, so active participation and contribution from developers all over the world can be achieved. However, free access to open-source code can be exploited for malicious purposes and potential vulnerabilities and security threats can be discovered.
11. Communication issues between the different expertise teams: For the effective realization of a smart city through IoT, multiple teams of people with different expertise, skills, knowledge, and experience must come together and cooperate effectively. However, these factors can cause issues in effective communication due to the lack of a common ground between them.
12. Stakeholders increase: With the introduction of the IoT in smart cities, different stakeholders are going to be involved, each one providing its own services. Examples include health monitoring, physical security, energy providing, and weather forecasting services. However, the lack of agreement between them in the correct deployment of strong security countermeasures can put multiple IoT networks at risk.
13. Coordination of humans and the smart city within IoT networks: The interaction between the smart city citizens and the smart city's IoT networks and cyber-physical systems must be correct, accurate, reliable, safe, and secure. Furthermore, detailed visual representation of processes and procedures and important information is important so smart citizens can have an accurate and real-time representation of the smart city's state. These procedures can prove to be difficult in coordination as multiple stakeholders, networks, hardware/software, and humans are involved.
14. Fault and fail-safe tolerance and mitigation: In smart city's IoT networks and especially critical infrastructures, both fault tolerance and safe failing are vital so fatal consequences are avoided in critical events such as environmental disaster or holistic power failure.

8.5.2 IoT in Smart Cities Threats and Attacks

The vulnerabilities presented in the previous section can be exploited to cause threats to the IoT and subsequently the smart city. We proceed by presenting an extended list of IoT threats. A summary of the IoT attacks is illustrated in Figure 8.3.

8.5.2.1 IoT Hardware Attacks

1. Side Channel Attacks: In a side channel attack, the adversary makes use of physical implementation information such as timing data, power consumption, electromagnetic leaks or even sounds which can be used to exploit the device further. Most of these attacks are based on statistical approaches but could also require more technical information such as knowledge of the operating system. A notable example of a side channel attack in IoT includes the attacker using the sound a 3D printer emits when creating an object to reconstruct the objects at a later stage without the need for their design [41].
2. Device Tampering Attack: The adversary gets direct physical access to the device and compromises it so they can most likely get access to the device's resources (e.g., memory) [42].
3. Battery Drainage Attacks: The adversary aims to exhaust the battery of a device by sending a burst amount of legitimate requests in order for the device not to be put in sleep or energy-saving mode and eventually drains its battery [43].
4. Hardware Trojan: The adversary maliciously modifies the circuitry of an integrated circuit. This malicious act can occur either at the design or the

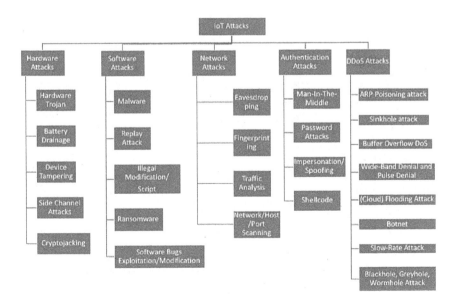

FIGURE 8.3 IoT attacks taxonomy.

fabrication stage and can result in the malfunctioning of integrated circuit during its runtime. A hardware Trojan can be maliciously installed at any phase of the very large-scale integration (VLSI) design life cycle. VLSI denotes the procedure of assembling an integrated circuit which includes bringing into one microchip millions of MOS transistors. A hardware Trojan can use radio emission to leak private and confidential information. More importantly, a hardware Trojan can be proved catastrophic to the microchip as it could disable or derange parts of it or even destroy it completely. A hardware Trojan can be activated under multiple and different circumstances as described in [44, 45, 46]. We briefly describe the most popular approaches below:

    a. Always On: Some types of hardware Trojan are triggered as soon as the integrated circuit starts operating. In such cases, the hardware Trojan's main aim is to constantly monitor the microchip's activity so various malicious acts can be performed. Some examples include leaking of information, bypass instruction, chip compromise, or overall system damage.

    b. Triggered: This type of hardware Trojan is activated when certain conditions regarding the microchip are met, either internal or external.

        i. Internal conditions can either be physical, logical, or timing. Physical has to do with the physical state of the chip such as temperature, pressure, stress, strain, or geo-location. Logical conditions have to do with logical conditions being met such as computational or computational activities regarding the chip. Finally, timing conditions have to do with the Trojan being activated after a certain time passes.

        ii. External conditions can be triggered either by user's input or an adversary's command sent.

5. Cryptojacking: This is defined as an adversary maliciously compromising a device and using it for cryptocurrency mining. Although IoT devices are lower in resources needed in mining such as memory and CPU power, the fact that they are so easy to be compromised makes them an ideal target for cybercriminals as they can affect a very large number of them with rather low resources [45, 47].

8.5.2.2 IoT Attacks on the Network

1. Eavesdropping: The adversary "listens" to the unauthorized information transmitted through a network communication [48]. An example in IoT networks would be the adversary eavesdropping on the packet transmissions of motion and light sensors so they know when a smart home's owner is inside or not.

2. Traffic Analysis: In contrast to eavesdropping, the attacker is interested in analyzing the communication patterns by passively monitoring them [48]. An example in IoT networks would be the adversary performing traffic analysis so they can understand what kind of sensors, actuators, and systems are deployed in a smart building.

3. Network/Host/Port Scanning: The adversary scans the network so they can gather more information regarding the services running, open ports, IP addresses, and so on. In IoT networks, an adversary can perform this set of malicious acts to identify open ports and their IP addresses and check what services/protocols these IoT devices run [49].
4. Fingerprinting: The adversary scans a specific host after it has identified it is active in the network to identify its operating system so they can investigate for potential operating system vulnerabilities [48].

8.5.2.3 IoT DDoS Attacks

A Distributed/Denial of Service (DDoS) attack can be volumetric or vulnerability exploitation-based or reflection-based. A volumetric DDoS attack consists of an attacker sending a massive volume of requests in an attempt to flood the target machine and make it unable to accept legitimate requests from normal users. In a vulnerability-exploitation DDoS attack, an attacker is taking advantage of a system or protocol or communication weakness to make the target machine unavailable to legitimate requests from normal users. In a reflection D/DoS attack, the target is responding with the received challenge. A DDoS attack can also occur in any of the TCP/IP layers. DDoS attacks mainly threaten the Availability Security Principle. Below, we briefly describe the most popular DDoS attacks across all layers.

DDoS attacks in IoT networks have a more critical impact compared to traditional ICT networks' operational capabilities of the device or the entire IoT network as the authors in [50, 51] and [52] effectively stated. Hence, we proceed with listing and explaining some of the most popular D/DoS attacks with a short description and their impact based on the TCP/IP protocol stack layer they operate as an extension to the previous D/DoS attacks mentioned in the previous section which can also be conducted in IoT environments. We proceed to describe a set of DDoS attacks that are quite popular in IoT networks. Beyond impacting data and resources, they can actually negatively affect the following.

1. Jamming: An attacker floods the physical medium with a burst amount of signals in an attempt to prevent legitimate packets from being transmitted successfully. Jamming can be constant where a large amount of radio signals are sent from the start and in a constant manner. Jamming can also be deceptive where the adversary falsely sends radio signals to make the target unavailable instead of the communication medium [53–57].
2. Radio Frequency (RF) Jamming: RF jamming is a special type of jamming DoS attack where the attacker sends RF signals so they can interfere with the communication among different devices, which leads to failure in communication with each other. This is a volumetric type of attack and can be conducted on the physical layer [58, 59].
3. Wide-Band Denial and Pulse Denial: The attacker blocks the RF in the IoT network completely leading to communication loss. This is a volumetric type of attack and can be conducted on the physical layer [50, 51].

4. Cloud Flooding Attack: The attacker sends a burst amount of singled-out requests and therefore exhausts the cloud resources because of which the cloud server is unable to respond to legitimate requests from other IoT devices within the network [50, 51].
5. Buffer Overflow DoS: The attacker sends a rather large chunk of data that ends up overwriting part of the actual application and therefore making it unavailable to legitimate users [50, 51].
6. Node Jamming in WSNs: The attacker aims to disrupt the wireless communication between nodes in a WSN by actively jamming the wireless channel [42, 60].
7. Sinkhole: The attacker compromises one of the nodes of the actual IoT network and proceeds with sending fake routing information to other nodes, falsely claiming to know the minimum distance path to the base station. The other nodes send the malicious node traffic which it either alters or drops [61–64].
8. ARP Exploitation Poisoning: The attacker performs MAC address spoofing so it can send ARP messages to a network. Hence, messages are redirected to him instead of the normal user. Therefore, the attacker can cause a DDoS attack by dropping the packets. In wireless networks, a node that wants to connect to a wireless network scans the environment to connect to an available network by sending Probe Requests. The access point responds, providing information about the network. In a Probe Request Flooding attack, the adversary sends a burst amount of probe requests to the access point to make it unavailable to legitimate requests [65].
9. MAC Authentication Request Flood: The attacker who has already performed a MAC spoofing attack to try and authenticate themselves to the access point by sending authentication requests. The attacker aims to flood the access point with authentication requests.
10. Blackhole: A compromised node that is supposed to forward packets received drops them [60, 66].
11. Greyhole: A compromised node that is supposed to forward packets received selectively drops them [60, 66].
12. Wormhole: A compromised node that is supposed to forward packets received to specific nodes, forwards them to the wrong node maliciously [67–68].
13. Flooding DDoS: In this type of attacks, the attacker sends a burst amount of either malformed, oversized or legitimate packets in an attempt to flood the target. Popular examples of such attacks include internet control message protocol (ICMP), user datagram protocol (UDP), transmission control protocol (TCP), hypertext transfer protocol (HTTP) DDoS attacks which use the respective protocols [52].
14. Low and Slow Rate DDoS: In contrast to flooding DDoS attacks, this is an attempt to make the target unavailable by sending a small amount of volume sent to the target [52]. Popular examples include Slowloris, RUDY, and slow read. We briefly describe these examples below:

a. Slowloris: The adversary successfully completes a TCP connection and sends a partial application layer protocol header request. Next, the adversary exploits the server's waiting request time for requests yet to be completed by sending a small subsequent packet just before the server drops the connection so it can remain open and available. As time is passing by, more connections are initiated with partial header requests that are never completed, thus making the target unresponsive and unavailable to legitimate requests.
b. RUDY/Tors Hammer: The adversary completes a TCP connection successfully and proceeds with sending a legitimate application-layer header. However, during this attack, the adversary instead of sending an impartial header sends a partial request body. Once again, as time is passing by, more connections are initiated with partial body requests that are never completed, thus making the target unresponsive and unavailable to legitimate requests. The presence of forms is required in order for such an attack to be successful.
c. Slow Read: The adversary sends a legitimate application layer protocol request to the target server and successfully starts receiving the response from the server but reads it at a very slow speed. The attacker reads the response too slowly, sometimes as slow as 1 byte at a time, thus preventing the server from rejecting the request connection.
d. Buffer Overflow: An adversary maliciously exploits an application's weak software security ending up overwriting memory adjacent to a buffer that should not have been modified under any circumstances. This results in the crash of the application or making it unstable while functioning. This particular attack can be executed in IoT devices with more ease as they are lower in resources than regular computer machines.
15. Botnets: A botnet is a set of devices, most of the time large in size, which have been previously infected with some type of malware that exploited their vulnerabilities and gave root privileges to the attacker. The attacker (in a botnet case, the botmaster) has compromised this machine and maliciously uses it to perform a malicious act, such as DDoS attack, spamming or clickfrauds [48].

8.5.2.4 IoT Attacks on Authentication

1. Impersonation/Spoofing: An adversary manages to successfully impersonate another entity's identity to falsely gain an advantage [48]. Examples of an impersonation attack include IP, GPS, DNS, and ARP spoofing. An impersonation/spoofing attack in IoT would consist of a smart meter being impersonated so a chosen victim pays the attacker's electricity bill.
2. Password Attacks: This set of attacks targets to achieve unauthorized access and possibly impersonation through maliciously bypassing password authentication mechanisms [48]. Examples of such attacks include brute

force attacks and dictionary attacks. Brute force attacks attempt to try all the possible combinations of passwords one by one taking a trial and error approach. Dictionary attacks, on the other hand, adopt a smarter approach as they use a large dictionary of meaningful words and popular or previously used passwords. IoT devices have been subject to such attacks as the Mirai incident has clearly demonstrated [69].

3. Shellcode: The adversary uses a small piece of code so they can exploit a vulnerability present in a software application, resulting in the attacker managing to remotely control the compromised machine [48]. In IoT networks, this can be part of a larger and more complex attack in an attempt to compromise the machine and include it in a botnet.
4. Man-In-The-Middle (MITM): In MITM, the adversary illegally intercepts a communication between two entities A and B, by pretending to be entity A to B and vice versa [48]. Furthermore, the adversary possibly alters the messages that are supposed to be exchanged between the two entities. In IoT networks, such a case would occur when an adversary would like to intercept confidential information between IoT devices such as a patient's medical data forwarded from a medical device such as their glucose monitor device to their personal doctor. Another example would consist of an adversary intercepting the communication between sensors monitoring a power plant and reporting its state back to the relevant server, aiming to alter the commands send back to the sensors and put the power plant's normal functioning at risk.

8.5.2.5 IoT Malicious Software

1. Malware: In an attempt to compromise a device and misuse it, an adversary can illegally insert a malicious software application. In IoT devices, it is easier compared to regular computer machines due to the former's weak or non-existent security countermeasures [48]. When malware is successfully installed in IoT devices, the results can be severe as they can start misbehaving, thus causing issues to people using them and can infect other IoT devices through propagation.
2. Spyware: Spyware is a special type of malware that spies on a device so they can steal personal and confidential information [48]. An example in IoT can be installation of a spyware for stealing a person's credit card information used from a smart fridge for automatic goods reordering.
3. Ransomware: Ransomware is another special type of malicious software that focuses on holding data and/or access to a device or system, thus having the data and the device's services hostage until the victim pays a ransom. Its extortion capabilities include encryption of data or the entire disk partition, their destruction or locking the user out of the device or system [48]. Although many IoT devices are unlikely to hold any important data due to their low memory capabilities, they can become unavailable for usage when a ransomware infects them by locking the user out. Such an event can be quite fatal if, for example, medical equipment in intensive care units is infected with ransomware and cannot serve the patients in need.

Security Challenges and Solutions in IoT Networks for the Smart Cities   179

4. Software Bugs Exploitation/Modification: IoT are considered parts of a cyber-physical system where based on digital commands/signals, it performs physical actions. Hence, exploitation of its software or a discovery of a bug can result in malfunctioning of the hardware and therefore misuse of the device. Some examples include the smart electric/water meter being unable to forward power/water to the home/building's premises after a potential software bug exploitation.
5. Illegal Modification/Script attack: The adversary illegally modified messages to be sent to a victim or inserts a malicious script. An illegal modification attack can also lead to privilege escalation [48].
6. Replay attack: The adversary illegally delays or repeats data messages. A replay attack can also lead to privilege escalation [48].

8.6 IoT ATTACKS COUNTERMEASURES

A plethora of different solutions and countermeasures have been proposed in the literature throughout the years for the accurate detection of various threats in IoT networks. Next, we present and describe a set of notable lightweight solutions, categorized on the threat they guard against.

8.6.1 COUNTERMEASURES ON IoT HARDWARE ATTACKS

In order for IoT hardware attacks [46] to be accurately detected and mitigated, proper countermeasures have to be implemented that aim to monitor hardware characteristics and accurately detect abnormal changes, always taking into consideration that IoT devices are low in resources. In this section, we summarize a set of notable studies against hardware attacks.

Affected by machine learning's popularity, the authors in [70] have used a popular supervised machine learning algorithm called support vector machines (SVM) to detect three different types of hardware Trojan netlists. These hardware trojans include a trigger circuit composed of either a combinational, sequential type, or no circuit at all.

For the SVM's training phase, seven different benchmarks obtained from Trust-HUB were used. The following features were extracted to train SVM: 1) the number of logic gates two levels away from the netlist inputs; 2) the number of logic levels to the nearest flip-flop input from the netlist; 3) the number of logic levels to the nearest flip-flop output from the netlist; 4) the minimum logic level from any primary input to the netlist; 5) the minimum logic level to any primary output from the netlist.

During the evaluation phase, the constructed SVM model is tested against unknown netlists. For the combination circuits, the SVM algorithm can detect all hardware Trojans (100% TP rate); however, it generates a 53.3% FP rate. For the sequential circuits, SVM once again achieves a 100% TP rate, but its false alarm rate remains high with a 32.7% FP rate. Finally, for the no-circuit hardware Trojans, the TP rate falls drastically with only 11.1% TP rate and an FP rate of 32.5%.

In contrast, the authors in [71] used three different types of electromagnetic leakage signals to detect hardware Trojans: 1) hardware Trojans that leak the cryptographic chip's key using electromagnetic emission, 2) hardware Trojans that obtain the key or data by using spectrum to form the electromagnetic leakage pass below the device's noise level, and 3) hardware Trojans that leak the key by encoding the time slot between the bytes.

The electromagnetic signal curve is extracted and the dimension of data is reduced using the Principal Component Analysis (PCA). Every electromagnetic signal curve collected is transformed as a feature vector, and the sampling points of the curve denotes the feature vector dimension, hence PCA was used to reduce its dimensionality. Next, the Mahalanobis distance was calculated between the chip and the groups constructed (Trojan free chip and template/target chips). Finally, based on the threshold decided, the chip was classified accordingly. The authors reported in their experimental evaluation results about their detection algorithm achieving a 91.93% accuracy and an average time consumption of only 0.042 s.

To mitigate and possibly prevent the network against hardware Trojans, there are other more practical and straightforward techniques that can be adopted as the authors in [72] stated. They proposed temperature analysis as an effective countermeasure against hardware Trojans. Specifically, they explained that the temperature is most likely to be increased and more heat will be generated if there is an additional integrated circuit on top of the original one. Hence, monitoring the temperature could potentially be used to prevent Trojans. However, they correctly argued that the temperature might be increased due to other non-malicious reasons so more research is needed. The authors also presented additional practical solutions proposed in previous studies, specifically from [46, 73, 74].

The authors in [46] proposed measuring the total current flow for each specific part of the integrated circuit. In the case of presence of hardware Trojans, the current flow for that specific part will be increased compared to the rest. However, as the authors in [72] correctly pointed out, hardware Trojans are very small in size and need a very small amount of current to function. As a result, they are unlikely to cause any significant changes in the current flow.

Another method for detecting hardware Trojans is the monitoring of delay time used by the integrated circuit as the authors in [73] argued. If a hardware Trojans is in place, then the integrated circuit will experience delays. This method is unaffected against the small size of the hardware Trojans. However, as the authors in [74] once again effectively pointed out, a comparison between legitimate and malicious delays of the integrated circuit must be made to create a baseline of legitimate behavior. Unfortunately, data regarding normal delays in integrated circuits might not always be available.

Finally, the authors in [74] proposed monitoring the power consumption to detect the presence of hardware Trojans. In the case of presence of a hardware Trojan in the chip, the electric consumption will be increased. However, this technique will not be able to precisely locate where the hardware Trojan is exactly placed, as the authors in [72] pointed out. In addition, as previously discussed, measuring power/current flow is unlikely to demonstrate any significant variations due to the hardware Trojan's small size.

Due to the potential dangers the globalization of manufacturing IoT devices poses, the protection of the systems-on-chip is more important than ever. Hence, the authors in [75] proposed RG-Secure: a multi-layer hardware Trojan protection framework for the IoT perception layer. RG-Secure brings together intellectual property trusted design strategy and scan-chain netlist feature analysis technology. The authors proposed a system focused on detecting RTL Trojans and gate-level Trojans. They used lightGMB, a distributed, lightweight gradient lifting algorithm. Their choice was based on the algorithm's capability for effectively processing high-dimensional circuit feature information, which resulted in accurate detection. For the Trojans' detection, they made usage of supply chain and Trojan features. The hardware Trojan features are the following:

1. GFi (Logic Gate Fan-ins): For each net, count the number of inputs in the logic gates that are two levels away.
2. FFi (FlipFlop Input): For each net, count the number of logic levels to the nearest flip-flop input.
3. FFo (FlipFlop Output): For each net, count the number of logic levels to the nearest flip-flop output.
4. PI (Primary Input): For each net, measure the minimum logic level from any primary input.
5. PO (Primary Output): For each net, measure the minimum logic level from any primary output.

More information regarding the hardware Trojan features is discussed in the authors' previous work in [76]. For the scan-chain features, every sequence element (e.g., D Flip-Flop) gives its place to a scan sequence element (e.g., Scan D Flip Flop). The output of the Scan D Flip-Flop is connected to the input interface of the next Scan D Flip-Flop to create a scan chain.

RG-Secure detection system operates in the following way: firstly, a gate-level netlist is accepted as an input. Next, the four classifications are initialized (Suspicious Trigger Module, Suspicious Observe Point, and Trojan net). Then, the netlist is classified based on the features selected. To improve the detection rate, scan-chain features are also used for all the netlists. Furthermore, conditional selection strategies are used to analyze suspicious netlists. Based on this set of procedures, each netlist is classified accordingly.

Their experimental results demonstrate that RG-Secure can successfully detect both register-transfer-level and gate-level hardware Trojans simultaneously. In detail, they report a 100% TP and a 6% FP rate, which indicates that their proposed solution is a promising approach to securing systems-on-chip during the design phase.

Highlighting the need for lightweight defense solutions to IoT hardware attacks, the authors in [77] have proposed a dynamic permutation defense scheme against hardware Trojans and side-channel attacks. In detail, they effectively discuss that computation chips will likely encompass a lightweight cryptographic algorithm to ensure confidentiality. Hence, they proposed dynamic permutation as a defense method.

When dynamic permutation is used, the order of the collected data from the sensor is changed. In that way, an adversary cannot execute a hardware Trojan with a predefined condition while the processor is processing a dynamically permuted message. The larger the number of permutation patterns allowed, the longer it takes for the attacker to succeed in obtaining the unit design details or even executing an attack. Furthermore, even if the adversary processes any information regarding the processing unit design, the dynamic nature of the permutation method still makes it harder for them to execute an attack or obtain the cryptokey. The cryptographic module of the processor validates the permutation pattern and requests a new one in case it is weak. Hence, the authors used permutation randomness of a 5-bit counter which resulted in a nearly flat accumulated partial guessing entropy with a subkey byte obtaining 2 for 7000 power traces.

While sophisticated and tailor-made solutions for all types of threats in IoT networks are recommended, more practical recommendations can prove exceptionally suitable for protecting them. Hence, the authors in [78] have focused on providing a set of effective practical recommendations for securing IoT devices based on their hardware-oriented procedures. The first recommendation consisted of performing full disk encryption to protect any data stored in the device. Due to resource constraints, IoT devices might have weak authentication mechanisms as demonstrated in Mirai. Therefore, IoT devices, which store confidential data, are easy to be compromised and cause information leakage.

The second countermeasure suggested is the usage of a cryptoprocessor that can store the encryption key and modify the boot loader so the system can verify its integrity. The cryptoprocessor stores the key after encrypting it and can only be used when the IoT device is in the sealing state. Hence, there is no need for password manual entry. Furthermore, in order for the system to be verified, it has to be ensured that no unauthorized software is in usage before the key is released during the boot process. Therefore, a trust process should involve hardware components. Next, we present and briefly describe the final two countermeasures proposed by the authors that are used to build this trust.

The third countermeasure involves using hardware that consists of the boot ROM that verifies the boot loader signature. Using a boot ROM that is able to successfully check the signature before execution denotes the prevention of an adversary maliciously changing the boot loader. Furthermore, if reverse engineering is attempted, the encryption key cannot be obtained since the secure boot ROM will not execute the boot loader because it has not been verified by a signature. As a result, the secret key is embedded in the hardware and once it is written will not be modified. The fourth and final countermeasure proposed was the usage of hardware that consists of an encrypted boot loader with a key which is also known to the boot ROM. Once again to prevent reverse engineering, the boot loader code can be encrypted.

Hardware characteristics could also be used for identity and trust management, data provenance, privacy, and data integrity in IoT devices as the authors in [79] pointed out. To ensure data provenance, data integrity, and identity management, they made usage of sensor physical unclonable function (PUF) technology.

A PUF denotes a physical object that gives a physical "digital fingerprint" output (response). This response can be used as a unique identifier based on a specific set of input and conditions (challenge). PUF objects consist of unique physical variations which are constructed randomly during the semiconductor manufacturing process. PUFs are usually placed in integrated circuits.

A sensor PUF enhances a regular PUF's functionality so that it offers authentication, unclonability, and verification of a sensed value through variations. These variations are provided by using two types of inputs. The first input is a physical quantity and the second is a regular binary challenge.

PUF sensors encompass a set of different characteristics. Firstly, for any given set of challenges and sensed quantities, the sensor PUF outputs the same response. Secondly, none of the challenge-quantity-response triple reveals any other information about other triples. Finally, there is no possible way for the manufacturer of the sensor PUF to predetermine the challenge-quantity-response mapping.

Using them, the authors correctly argued that adversaries can no longer spoof measurements. PUF sensors prevent them from interfering with the analog signals that are responsible for transferring the sensor element to the embedded microprocessor. Furthermore, it ensures data integrity as it produces the same response for a given sensed quantity.

Regarding identity management, the usage of sensor PUFs can prove very beneficial. A sensor PUF provides the device with a unique ID. Therefore, it automatically gains its own distinct identity. This identity can be validated by using multiple and different challenge-response pairs.

To provide effective trust management and monitoring of applications' integrity, they used hardware performance counters. As the authors in [79, 80] stated, the root of trust begins at the hardware level. Using performance counters, malicious tampering as well as rootkits [81] can be effectively detected. Hardware performance counters are responsible for monitoring specific events during the execution of a program [82]. In detail, the counters start operating through the operating system as soon as the program starts. The events can be monitored either periodically or after the program finishes its execution. Through the usage of the collected events, a model can be generated and used to monitor the software's integrity.

Similarly, highlighting the constrained resources that IoT devices have, their physical access to adversaries, and the risk of their cryptographic keys being exposed, the authors in [83] proposed hardware security measures to ensure the protection of the cryptographic keys and data of IoT devices using web services that use the CoAP) and RESTful protocols.

To secure the cryprographic keys in IoT devices, the authors proposed a secure element (a dedicated chip connected to the microcontroller of the IoT constrained device) responsible for carrying out all the cryptographic operations which are going to be isolated from the firmware. Hence, the probability of a tampering or probing attack successfully occurring is minimized. The private keys generated are protected and not exposed to the application's memory. On the other side, of the resource server, both the cryptographic keys and the application data are secured in a trusted execution environment (a secure area inside the microprocessor). To achieve this, they used the Intel Software Guard Extensions technology which acts as a sandbox and protects code and data from various parts of the system such as the operating system, BIOS, drivers, and firmware. In that way, all code and data are secured from potential external malicious acts. As a result, the cryptographic keys used are strongly isolated and protected. To evaluate their proposed system, the authors measured overall processing time of the messages and the energy required for establishing the session. The results clearly show that no overhead is caused on the resource-constrained device used (Microchip ATECC608A).

8.6.2 Countermeasures on IoT Attacks on the Network and DDoS Attacks

Different types of intelligent techniques have been proposed to detect various DDoS and other network attacks on IoT networks. It is evident that machine learning and A.I. algorithms are strongly in favor because of their high accuracy when deployed in similar cyberattacks in regular ICT networks in the past. However, as already argued, IoT networks are low in resources and therefore unlikely to host such heavy algorithms. In this section, we present and describe notable studies in intrusion detection using machine learning and A.I. techniques as well as consider the low-in-resources nature of IoT networks, therefore proposing various methods to reduce their complexity.

Highlighting the need for lightweight detection, the authors in [84] developed an intrusion detection system (IDS) for IoT devices aiming to detect multiple types of DDoS attacks. They have chosen various host-based features regarding the Network Time Protocol (NTP) servers. NTP is a network protocol for clock synchronization between computer systems over packet-switched, variable-latency data networks. The host features chosen were delay time, response delay time, and offset clock. For the detection of DDoS attack, the mean and standard deviation are calculated and the central limit theorem is used to differentiate between legitimate and malicious DDoS traffic. In their experiments, the attack duration lasted for approximately 20 minutes. In their results, they report the TP rate being 100% and the FP rate being 16.66%.

Arguing about the impact wormhole DDoS attacks can have on IoT, the authors in [85] evaluated different machine learning approaches to detect such an attack. In detail, the authors proposed three centralized approaches; the first one is using the decision tree algorithm, the second one being K-means clustering, and the third one being a combination of the two. For the training phase of the decision tree, the safe distance between two neighborhood routers is used. The decision tree algorithm is trained through learning the safe distance between two neighborhood routers, and ultimately devices are attacked or are safe to be paired. K-means clustering is responsible for identifying safe zones so the FP rate can be reduced. During the experimental evaluation, the authors used different network topologies (mesh, ring, star) and different numbers of nodes deployed (10, 50, 100, 200). In their results, they report the TP rate for K-means being 70–93%, for decision tree 71–80%, and the hybrid approach being 71–75%.

Similarly, the authors in [86] justified that WSN networks require lightweight systems for detecting blackhole and wormhole attacks. Therefore, they captured network features, including the number of packets dropped, data packets received, and the number of packets forwarded, and logistic regression to detect these two attacks. In their results, for the wormhole, the TP rate was 75% with no FP rate being generated. For the blackhole attacks, the TP rate was 100%, again with no FP rate generated. Furthermore, the authors constructed an attack scenario where both attacks were conducted simultaneously with the TP rate being 91%.

Due to the severe impact of not only blackhole but also flooding attacks on IoT networks, the authors in [87] used machine learning detection techniques to protect them. These algorithms include naïve Bayes, decision trees, random trees, and Bayesian networks. For blackhole attacks, naive Bayes had a TP rate of 75.7895%, decision tree 98.9183%, random tree 74.7596%, and Bayes networks 77.1635%. For the flooding experiments: naïve Bayes 77.381%, decision tree 72.6316%, random tree 72.6316%, and Bayes networks 81.0526%.

The authors in [88] also used logistic regression, highlighting its suitability for lightweight detection approaches, to detect blackhole and jamming attacks in WSNs. In their experiments, they simulated each attack scenario for 15 minutes with it being repeated 5 times on each. In their results, the authors reported that the TP rate for constant jamming and blackhole attacks was 95% whereas for the combination of blackhole and random jamming, it was 60%. With regards to the FP rate, the results were approximately 0.25% and for blackhole and random jamming 0.25%.

Using the KDD99 dataset, the authors in [89] to detect all of the attacks included all of those that can be applied in IoT networks. Since the dataset consists of a great number of cyberattacks, a large set of features is needed. However, since IoT devices are low in resources, the authors used Principal Components Analysis (PCA) so the number of features is reduced. For detecting the attacks, two algorithms were evaluated: KNN and softmax regression. The dataset consisted of 490,000 instances, from which 60% were used for training and the remaining 40% for testing. Regarding the feature reduction procedure, three types of results were generated, one with three, one with six, and one with ten features. For the evaluation of the KNN and softmax regression algorithm, three and six features were used achieving an accuracy of 85.24% and 85.19%, respectively. Similarly, softmax regression achieved an accuracy rate of 84.50% and 84.40%, respectively. Regarding complexity, softmax regression has a $O(m*n)$, where n is the number of features and m is the amount of information which comes with former iterations (between 5 and 20). Therefore, the complexity is $O(1)$. For the KNN, the training phase is $O(n*\log n)$. For the testing phase, it is $O(n+n*\log n)$. Hence, KNN has a significantly higher complexity for both phases. Softmax regression's FP rate generated was 5% for both three and six features. No information about the FP rate of KNN is given. During testing, 144,021 instances have been classified. Softmax regression required 5.72 seconds while KNN required 507.31, a result that is justified by the significant complexity difference.

The authors in [90] effectively highlighted that IoT networks are high in network size and heterogeneity, therefore their optimized management and monitoring are essential. Hence, the deployment of software defined networking (SDN) is a beneficial solution to manage the issues stated. As a result, they proposed an IoT IDS (called SofThings) to detect DDoS attacks, specifically TCP and ICMP Flooding, in SDN-based IoT networks. Their proposed system consisted of a cluster-SDN controller installed in every cluster and a master-SDN controller at the top of the IDS hierarchy. The cluster controllers notify the master controller if any changes are detected in traffic. The master SDN controller is responsible for classifying the forwarded traffic as malicious or normal. Regarding the detection algorithm, the authors evaluated both linear and non-linear (radial basis function [RBF] kernel) SVM. The authors reported choosing to monitor various network features, including the number of sent requests, number of failed authentication attempts, source of requests, bandwidth consumption device usage at different periods, and so on. In their experimental evaluation, they have constructed three scenarios to evaluate their proposed method's accuracy. In the first scenario, a TCP flooding was conducted; in the second, an ICMP flooding was conducted; and in the third, a DDoS attack was conducted. For the first scenario, linear SVM achieved 94% precision and 92% TP rate while non-linear SVM achieved 98% precision and 97% TP rate. For the second scenario, linear SVM achieved 92% precision and 88% TP rate while non-linear SVM achieved 97% precision and 96% TP rate. Finally, for the third scenario, linear SVM achieved 94% precision and 93% TP rate while non-linear SVM achieved 98% precision and 95% TP rate.

Once again, the importance of being lightweight for any intrusion detection solutions in IoT networks is presented, as the authors in [91] constructed a dual two-tier IDS, the first tier for feature reduction and the second tier for classification of the incoming traffic to detect attacks from the NSL-KDD dataset. For the feature reduction part, PCA and linear discriminate analysis were used and for the classification part, naive Bayes and KNN were used. For the training phase, two sets were used: the first consisted of 251,92 instances and the 125,973. After the feature reduction process, the number of features was reduced from 41 to 35. The computational complexity of naive Bayes is $O(e*f)$ where e is the count of samples in the dataset and f is the number of features and for KNN is $O(\log n)$. In their results, the authors reported a TP rate of 84.86% and an FP rate of 4.86%. The authors also performed multi-class classification with their results being for normal 94.43%, for probe 87.32%, for DoS 88.20%, for U2R 70.15%, and for R2L 42%.

Highlighting the difficulty in detecting application-layer DDoS attacks, the authors in [92] proposed a lightweight IDS for detecting application-layer DDoS attacks for the smart home IoT network. They constructed a lightweight algorithm consisting of forecasting and chaos theory to detect application layer flooding and slow-rate attacks as well as correctly identifying flash crowd legitimate traffic. In their results, they report their IDS TP rates ranging from 87.5% to 100% and the maximum FP rate at 12.5%.

8.6.3 COUNTERMEASURES ON IoT AUTHENTICATION AND ILLEGAL MODIFICATION ATTACKS

In general, the literature highlights the weak or no presence at all of security countermeasures in IoT devices due to resource constraints. Therefore, it is important for lightweight solutions to be developed that are also highly successful despite their simplicity compared to equivalent solutions designed for systems that are rich in various resources such as memory, CPU power, and storage space.

This view is shared by the authors in [93], who have also discussed the need for two-factor authentication because many IoT devices will be deployed in open and public spaces, highlighting the necessity of the authentication protocol to be robust even if the device falls into the hands of adversaries.

As a result, the authors proposed a two-factor authentication scheme that integrates a second defense layer where PUFs are used in addition to a first defense layer of password authentication. PUFs denote the random physical variations found in an integrated circuit's microstructure during manufacturing. These variations are unique and cannot be replicated or derived, hence, nearly impossible to be either cloned or predicted. As a result, a PUF can make usage of its internal structure to generate a function that cannot be replicated by anyone and therefore proved to be a good authentication mechanism.

Their proposed scheme consisted of two phases: setup and authentication. During the setup phase (which must be carried out on a secure channel), the following steps are executed.

1. The IoT device initiates its communication with the server by sending its identity and a registration request.
2. The server receives the request and proceeds with generating a challenge for its next interactions with the IoT device as well as a set of new challenges which will be used for resynchronization with the IoT device and send them to it.
3. Upon receiving the challenges, the IoT device constructs the PUF and sends it back to the server.
4. Then, the server generates a one-time alias identity (using its master key and the physical unclonable functions of the IoT device concatenation) and a secret key to use it as a first authentication factor in the process of ensuring that the IoT device is legitimate.
5. Furthermore, the server carries on with generating multiple fake identities and synchronization key pairs that are unique to the IoT device.
6. As a final step, the server stores the one-time alias identity along with the secret key, the challenge response pair, the challenge and the set of challenges generated, and the fake identity and synchronization key pairs for the specific IoT device.

During the authentication phase, the following steps are executed.

1. When the IoT device intends to communicate with the server, it selects the one-time alias identity and proceeds with generating a random number by performing the XOR operation of it and the secret key. In the end, the IoT device constructs a request message which consist of the two and sends it to the server.
2. The server receives the authentication message and starts by locating the one-time alias identity in its database. Next, it reads the challenge response pair and the secret key and proceeds with generating a nonce by performing the XOR operating between the secret key and the nonce. It then computes a key-hash response and stores all this information on a message and sends it back to the IoT device.
3. When the IoT device receives the message from the server, it extracts the PUF output and then proceeds with computing and checking the key-hash response. If it is not valid, then the IoT device stops executing the protocol. If it is valid, the IoT device authenticates the server and decodes the

nonce, and gets the key-element and heaper data. It proceeds with calculating the key-hash response and updates the challenge and the PUF output, hash response, one-time alias identity, and the session key. Finally, the IoT device forms a message with the new PUR output, key-hash response, and heaper data, and sends it back to the server.
4. Finally, when the server receives the message, it starts by computing and decoding the heaper data to obtain the key element. Then, it proceeds by verifying the key-hash response. If the verification is successful, then the IoT device is successfully authenticated by the server, with the latter calculating the session key. Next, it computes the new challenge and decodes the new PUF output and updates the alias identity and the secret key and stores them for future interaction with the IoT device. If the server cannot authenticate the IoT device, then it tries to interact with the device again by using an unused pair of fake identity and synchronization key with both ends deleting it in the end.

Through an extended series of evaluations, the authors have proved their proposed scheme to demonstrate resilience to masquerade attacks and password attacks and encompass anonymity and untraceability. In addition, through their experiments, they clearly proved their authentication scheme not to be computationally heavy as all of the computations on the IoT device side take 2.9 ms or less and on the server side 3.34 ms or less.

The authors in [94] have designed a lightweight authentication/encryption scheme for IoT devices which consist of ARM processors based on LEA-128-CTR encryption and Chaskey message authentication code (MAC) algorithms. Chaskey algorithm is ideal for 32-bit microcontrollers as it includes key scheduling in its functioning and only uses ARX operations. Furthermore, it only uses 7.0 cycles/byte for long (128 byte) and 10.6 cycles/byte for short (16 byte) messages. Its implementation requires only 402 bytes of ROM to be successfully executed. LEA-128-CTR encryption can operate on 32-bit microcontrollers and it is secure against all types of cell-block ciphers as its supported by its inventors. It also consists of ARX operations for 32-bit words. Using these two algorithms, the authors proposed a single instruction multiple data (SIMD)-structure authentication protocol between the device and its verifier. The authentication protocol consists of the following steps:

- The device initiates the interaction with the identifier by calculating the MAC code of the pre-shared key using its ID and the timestamp and encrypting additional authentication data (e.g., fingerprint, iris, and so on) using the same pre-shared key and sends them along with its ID and the timestamp.
- The server upon receiving the incoming data from the device decrypts the additional authentication data and the timestamp using the pre-shared key, verifies the MAC code, and proceeds with generating a session key and a nonce verifier. It then proceeds with encrypting its ID, session key, and nonce verifier. Finally, it calculates the MAC of these three pieces of information and sends it along with the encrypted data calculated back to the device.

- The device proceeds with decrypting the received data using the pre-shared key, verifies the MAC code, and encrypts the nonce value generated by the verifier subtracting 1 and sends the result value back to the verifier.
- Finally, the verifier checks the value received to fully evaluate the identity of the device and the authentication procedure is completed successfully.

Through their experimental evaluation, the authors sufficiently prove their proposed authentication protocol to be secure and robust against eavesdropping, replay, and illegal modification attacks. Furthermore, they provide results regarding its hardware requirements with their proposed solution only requiring 42.354 ms for Non-SIMD and 22.770 for SIMD structures at most, when the message is 1 megabyte.

Since the majority of IoT networks will consist of wireless communications, the authors in [95] have proposed a spoofing detection technique using Q-learning and Dyna-Q, which achieve the optimal detection threshold using reinforcement learning on the physical layer. In physical layer authentication schemes, basic characteristics are used for detection of malicious behavior such as received signal strengths (RSSs), channel impulse responses, RSS indicators (RSSIs), channel state information, and channel frequency responses where static thresholds are defined; when any of them are surpassed, an alarm is raised. However, static thresholds are unlikely to work efficiently and are likely to raise a high number of false alarms.

Therefore, the authors highlighted the need for a dynamic threshold. To achieve this, they have applied two reinforcement learning algorithms, Q-learning and Dyna-Q. Initially, they used a zero-sum authentication game to model the authentication procedure, which consists of a spoofer and the receiver. They derived the Nash Equilibrium in the static spoofing detection game based on the channel frequency responses and proved its uniqueness. Next, they developed an authentication methods based on Q-learning and Dyna-Q for a dynamic radio environment. Their results show a significant improvement in the error rate of the proposed scheme to just 5%, whereas with the static threshold the error rate was 14%.

8.6.4 COUNTERMEASURES AGAINST IoT MALICIOUS SOFTWARE

One of the critical challenges to consider when detecting malware is not only attempting to mitigate its infection on a device, but also to isolate it and prevent it from being propagated to the rest of the network. Malware propagation is an important issue, especially in networks that consists a considerable amount of devices connected that can communicate with each other directly without the involvement of a server, such as any IoT network in smart cities. In this section, we review notable studies that focus on malware detection and propagation prevention lightweight techniques and approaches in IoT networks.

Based on the justifications provided, the authors in [96] proposed a low computational overhead to detect malware in IoT networks. They also constructed a framework that combines multiple and different types of malware that propagate in the network along with their consequences on its performance so they could come up with an optimized solution for malware repression.

For the proposed malware detector, called HaRM, using HPC traces, 44 different diverse microarchitectural events were constructed and used as features. To identify the most important ones, the authors have used Correlation Attribute Evaluation, which calculates the Pearson correlation coefficient between each feature and its class using the covariance between them. As a result, the final set consisted of only eight different features.

The dataset used for the experimental evaluation consisted of 3000 benign and malware executables. Examples of benign applications included MiBench [97] and SPEC2006 [98], Linux system programs, browsers, text editors, and word processors. Examples of malware executables were collected from popular websites such as virustotal.com and classified on virushare.com and included Linux ELFs, python scripts, perl scripts, and bash scripts. The malware files were categorized into Backdoor (452 executables), Rootkits (350 executables), Viruses (650 executables), and Trojans (1169 executables).

The chosen ML algorithm was OneR with the training and evaluation procedure followed being a 10-k fold validation. OneR was also compared with other popular ML algorithms, including multi-layer perceptron (MLP), logistic regression, and JRip using 2, 4, and 8 HPC features. OneR achieved a total accuracy rate ranging from 93.03% to 94.7% with an average rate of 92.21%, only surpassed by MLP. However, as the authors effectively highlight, MLP creates a considerable computational overhead, specifically power consumption and latency (time to detect a malware) as opposed to OneR. Regarding power consumption, OneR requires only 25% of power consumption than MLP, and 30% lower compared to JRip or Logistic regression implementations. OneR also has the least latency compared to the rest of ML algorithms with only 10 nanoseconds.

Furthermore, the authors used the outputs of OneR to generate probabilistic outputs about the infection state to model the malware propagation phase that is unfortunately not detected by the classifier. They tested their propagation model in a 20-node IoT network deployed in a 5×5 squared meters area. Their proposed method achieves a throughput of nearly double the amount compared to IoT network without any defense, and up to 100% higher compared to other heuristic approaches.

Considering the efficiency and potentially high accuracy of machine learning algorithms and the secure and reliable communication capabilities of blockchain, the authors in [99] proposed a malware detection system that uses both of these technologies for Android IoT devices. Their proposed detection scheme consisted of three stages: clustering, classification, and blockchain. Features regarding request and run-time permissions as well as system and API calls were used for constructing the machine learning classifier model. We give a brief description of the overall detection system later.

The first step consists of the usage clustering algorithm for feature selection. The clustering algorithm calculates the weights for each feature set, optimizes them, and eliminates the unnecessary ones. The final optimized set of features is used from the classifier for effective model construction. They correctly highlight the success of tree-based algorithms in malware detection demonstrated in the literature and, therefore, used the decision tree combined with naive Bayes algorithm. A decision tree was constructed in detail based on the training set provided. All the attributes in the constructed tree are collected and the subset of the most important ones is found based on their influence in classifying an instance. For this set of attributes, their conditional probability is calculated for each class to be used in the naive Bayes algorithm. Finally, the authors used blockchain in the last stage of their proposed detection method. Blockchain is used to store the malware features and all their relevant information (e.g., name, family type, permission feature, sensitive behavior, APK feature, and transaction hash). In addition, through providing this data, the malware detection can be carried out faster and more accurately.

To evaluate the detection system, the authors constructed their own android malware dataset which consisted of 192 benign and 5560 malware apps, which were collected from Google Play and Chinese App stores. In their experimental evaluation, their improved naive Bayes classification achieves an overall 98% accuracy.

Similarly, the authors in [100] have also focused on Android malware detection in IoT networks. As they effectively discuss, Android mobile systems interact frequently with smart devices and appliances in various IoT networks, such as the smart home. However, Android is increasingly popular and open-source, which makes it an ideal target for attackers in developing malicious apps to control smart home devices through their apps installation in mobile systems. Examples of such potential cases include such malicious apps that are masked as legitimate and are used by the attacker to control smart devices/appliances in a smart home without the user being aware of the same. Hence, a static event-ware Android malware detection algorithm was proposed which focuses on protecting users from installed malicious applications on smart IoT devices called EveDroid. EveDroid aims to check all applications to be downloaded from relevant app stores are safe and secure and falsely pose as benign executables.

EveDroid consists of two phases: training and online detection. During training, EveDroid uses behavioral features taken from multiple APK files, which are labeled as either benign or malicious. The training phase consists of three stages: the call graph extraction, event group building, and model construction. During the call graph extraction, the call graphs from all the APK files are extracted using static analysis tools, removing the repetitive API calls. Then the event group building constructs the event groups for the APK files using event subgraph traverse, API calls encoding, and clustering. Lastly, the clusters formed are forwarded to the classifier, a neural network which uses the leaky ReLU activation function and Adam optimization algorithm.

When the neural network is fully trained, the online detection phase commences. Once again, call graph extraction and event group building are conducted to extract the event groups from the unknown app to be classified. The extracted values are forwarded to the neural network. The model output elements that represent the probability that the app is classified as benign or malicious and the highest probability classifies the app as either. For the neural network to be effectively trained and evaluated, four different datasets were constructed. In detail, 109,56 benign samples in 2014 from PlayDrone.com and 4000 new apps in 2018 from PlayStore.com and a total of 28,848 malicious samples from VirusShare.com were collected to construct four different datasets in the following years: 2013, 2014, 2017, and 2018.

Based on the results, EveDroid performs exceptionally well across all four of them, with a TP rate ranging from 94.7% to 99.8%. EveDroid is also compared with other similar malware detection algorithms and machine learning techniques (SVM, Naive Bayes, and Decision Tree) and is shown to outperform all of them.

Taking into consideration that IoT malware can spread fast and impact large-scale networks, the authors in [101] proposed EDIMA, a malware detection system that can be deployed at the user access gateway and can detect malicious activities of a malware. Incoming traffic samples are collected and the relevant features are extracted from these traffic samples collected. The features extracted are: number of unique destination IP addresses and their packet number (maximum, minimum, and mean). The extracted values for this set of features are forwarded to an ML classifier. The authors evaluated three different algorithms: random forest, KNN, and Gaussian naïve Bayes. Whenever a new malware is detected, the classifier needs to be retrained and compared to the old model. If no significant differences are found, the classifier is not replaced with a new version of it. In addition, an optional final step, which is ideal for networks that consist of large numbers of IoT devices includes the introduction of a sub-sampling module that forwards only a part of the IoT traffic in an attempt to reduce the overhead.

For the experimental evaluation, the authors constructed a network consisting of various IoT devices, a laptop, an Android smartphone, and a wireless access gateway to collect the traffic generated. To evaluate the ML classifiers, 60 15-minute traffic sessions were collected containing both benign and malicious traffic, divided in a ratio of 70:30. From the results obtained, the most successful algorithm is proved to be KNN with a 94.44% accuracy compared to random forest (88.80%) and Gaussian naive Bayes (77.78%).

In contrast to traditional malware detection approaches, the authors have proposed detecting malicious behavior in IoT networks using image recognition. They justified that an easier way to classify malware is analyzing its image rather than extracting a large number of features. In detail, the malware binary file can be reconstructed as an 8-bit sequence and then converted into a grey-scale image (with values ranging from 0 to 255). The construed image is forwarded to suitable ML algorithms for image classification, with one of the most popular examples being the Convolutional Neural Network (CNN). A conversion of a malware binary to image is a relatively lightweight process as it only needs the input vectors, in our case 8-bit vectors. For the classifier to be lighter, a two-layer shallow CNN was constructed with 5000 iterations, a 32-value training batch size and a 0.0001 learning rate. For evaluating the constructed CNN model, the authors used the Mirai and Gafgyt malware families and a 5-fold hold-out validation approach. In their results, they report a 93.33% TP rate and a 5.33% FP rate.

8.7 DISCUSSION AND FUTURE DIRECTIONS

Undoubtedly, the literature considers IoT security a multi-scale and multi-dimensional problem that is of high importance. As discussed throughout this chapter, a plethora of different attacks can be conducted in IoT networks that can have not only an impact on the network, individual devices, and data exchanged but also on human lives. Therefore, it is essential for us to have countermeasures and solutions proposed, which are intelligent and sophisticated in their functioning. On the other hand, it is of equal importance for these intelligent and sophisticated defense mechanisms to be as lightweight as possible. Their computational complexity and overall model construction should not be high in hardware resources so any disruption to the individual devices or the overall network's functioning can be avoided.

In the field of intrusion detection, various popular machine learning have been proposed. Examples include tree-based techniques such as decision and random trees, Bayesian techniques such as naive Bayesian and Bayesian networks, clustering such as K-means and estimation techniques such as logistic and softmax regression algorithms.

Based on the studies presented, we observe that machine learning and A.I. techniques are particularly effective against multiple types of DDoS and network attacks as the overall TP rate ranges from 70% to 100% and the FP rate from 0% to 16.66%.

Regarding network DDoS attacks such as blackhole, greyhole, and wormhole, as well as physical jamming, we observe that logistic regression, K-means clustering as well as decision trees have been proved particularly effective. In particular, logistic regression proved to achieve perfect detection for blackhole attacks and 75% of greyhole attacks in [86]. Also, in both cases, no FP rates were generated. Logistic regression was also effective in [88] where various types of jamming and blackhole were used where the detection rates were very high (95%) and the FP rates very low (0.25%). This proves that logistic regression is highly effective, possibly due to its approximating nature. Other popular machine learning algorithms have been used for the detection of such attacks, for example, K-means and decision trees with similar success as the authors in [85] reported. Based on the results provided, K-means achieves a better detection with a minimum difference of 13%. In contrast, authors in [87], who aimed to detect blackhole, wormhole, and flooding attacks, report decision tree algorithm achieving the highest TP rate compared to random trees and Bayesian networks. On the other hand, for the flooding attacks, decision tree achieves the lowest TP rate in contrast to the other two algorithms.

Furthermore, studies that aimed to detect flooding attacks also produced satisfactory results, always considering techniques that are as lightweight as possible as the authors in [90] used such as linear SVM and the authors in [84] used such as statistical measures. Their effectiveness was evident as they achieved a TP rate ranging from 92% to 100%. However, the study in [84] consisted of a rather high FP rate with 16.66%; hence, effective tuning of the statistical thresholds assigned should be done in the future.

Next, the authors in [92] pointed out that application-layer DDoS attacks are of great importance to be considered in IoT networks. Their stealthy nature and great similarity to legitimate behavior can deceive security countermeasures in place and significantly impact the IoT network and its devices. Hence, more work should be focused on DDoS attacks in the application layer. Taking into consideration smart home's resource-constraints nature, they constructed a lightweight algorithm and compared its results to popular machine learning algorithms, arguing that high accuracy rates can be achieved without necessarily deploying high complexity algorithms.

Finally, the studies in [89, 91] have considered a plethora of different network attacks using the KDD99 and NSL-KDD datasets. To detect multiple and different types of such attacks, effective feature reduction should be conducted, an issue which the authors have evidently considered before training their machine learning algorithms. The former used PCA and the latter component analysis and linear discriminate analysis to reduce the number of features and therefore the model complexity and training time. The results in both cases do not seem to be heavily impacted as they achieve a TP rate of at least 85%.

The research conducted in network intrusion is vast and with great success. However, there is always room for improvement. Although the authors in all the studies presented earlier have focused on proposing lightweight machine learning approaches, new algorithms could be developed using statistical, probabilistic, and mathematical models. For DDoS attacks in particular, forecasting methods could be considered such as the study conducted in [92]. Furthermore, more effort should be put in dimensionality and feature reduction, especially multiple attacks from different layers of the TCP/IP such as bio-inspired algorithms as described in [102]. In addition, new features could be considered for IoT device's hardware characteristics such as memory usage, power consumption, CPU power, and overall temperature. In cases where the IoT device has been compromised and used as part of a botnet, its resources are likely to be exhausted so monitoring such characteristics can be proved particularly effective in detecting such malicious behavior.

In addition, the studies presented overfocus on detecting DDoS attacks but do not consider alerting for compromised nodes/botnets before the actual DDoS phase. If the detection algorithm uses IP addresses to detect DDoS attacks, then the compromised nodes can easily change their IP address to evade detection. Furthermore, botnets can have hibernation stages where they conduct the DDoS attack and then put in sleep mode to once again avoid detection, which should also be investigated as a scenario so that effective solutions can be developed.

Another limitation observed is that none of the presented studies consider any DDoS attack on the application layer. Application-layer DDoS attacks are particularly effective due to their stealthiness and great similarity to legitimate traffic. Finally, as machine learning is heavily used as a defense mechanism, adversarial machine learning should be taken into consideration. Adversarial machine learning consists of malicious acts performed by adversaries to mislead machine learning defense mechanisms and being eventually bypassed.

Regarding the countermeasures proposed for the effective detection and mitigation of malware, we can once again observe that machine learning algorithms and techniques are in favor due to their robustness and high accuracy. The set of studies presented earlier has also taken into consideration that IoT networks that mainly consist of resource-constrained devices, and any countermeasures proposed should be as lightweight as possible.

By examining the results, we observe that the majority of the machine learning algorithms used cannot be considered too heavy in resources, with examples being OneR used by [96]. A combination of clustering, decision tree, and naive Bayes proposed by [99], MLP neural networks [100], and KNN used by [101] were discussed. In addition, the authors in [103] considered that the usage of a CNN neural network is likely to negatively impact IoT devices and they, therefore, proposed a novel approach in transforming the malware binary to an image and constructing a two-layer shallow CNN.

Based on the results provided, we observe that machine learning is proved to be particularly effective in detecting various types of malware such as Backdoors, Rootkits, Viruses, and Trojans (with famous examples being Mirai and Gafgyt malware, with the TP rates being at least 92%).

Results from the literature are more than reassuring and machine learning and A.I. techniques will continue being heavily used. In the future, it is more than evident that more sophisticated malware is going to target IoT devices due to their heavy usage in different types of networks and their easy-to-compromise nature. Hence, new and more sophisticated machine learning algorithms should be explored, always considering making any necessary changes in reducing their complexity and hardware resources. Possible approaches that could be adopted include dimensionality reduction techniques such as PCA and various bio-inspired algorithms as it was described in [102].

Furthermore, it is more than evident that IoT ransomware is going to be a trend in the near future. Ransomware is a special type of malware that must be studied closely and individually so its specific malicious behavior can be effectively detected and prevented before the data is encrypted or user lockout is performed. Similarly, cryptojacking malware is a special category that relevant attention should be paid as it exhausts the resources of IoT devices.

An interesting approach that could be explored as it was presented in the DDoS and network attacks suggestions include the monitoring of hardware characteristics to detect malware, including memory usage, CPU power, temperature, and voltage used. The infection of an IoT device from a malware is likely to cause an increase in the overall usage of these hardware characteristics; hence, they could be used as alternative features in malware detection. Similarly to the DDoS and network attack argument, adversarial machine learning should be considered and effective countermeasures should be deployed.

Studies that proposed countermeasures against hardware attacks in IoT consisted of multiple and different types of approaches. Once again machine learning techniques have been frequently chosen, in particular for hardware Trojans. However, in contrast to the studies for DDoS and network attacks, other techniques have proved to be more effective. This was clearly highlighted when comparing the studies from [70, 71, 77]. This clearly demonstrates that although machine learning's intelligent approach and general relative success in multiple and different problems in computer science and engineering is overwhelming, other methods should be considered. Furthermore, beyond accuracy, the other techniques (electromagnetic leakage modeling and distance measures and permutation algorithm) are more lightweight compared to the SVM algorithm and therefore more suitable for IoT devices. In addition, based on [78], considering practical solutions that are directly dependent on the hardware installed can improve the hardware security of IoT significantly. Such countermeasures are highly practical, easy to be deployed at no additional costs, and do not require any additional allocation of resources from the device.

Future directions in the hardware security of IoT should follow multiple pathways. Firstly, more focus should be placed in detecting not only hardware Trojans and side channel attacks, but also battery drainage and cryptojacking attacks, as well as software bug exploitation and physical tampering. All of the attacks mentioned can be executed with minimum resources and their sophistication cannot be considered high. They can potentially pose a serious threat and might potentially proved catastrophic, especially in critical infrastructure IoT networks.

For battery energy attacks, forecasting and prediction methods could be applied so an effective baseline of the device's energy consumption can be constructed and alert in case the current consumption exceeds this baseline by far. For cryptojacking attacks, along with battery levels and energy consumption, other hardware characteristics can be monitored such as memory usage and CPU power.

With regards to software bugs, various approaches could be considered including dynamic programming and extensive code testing before its deployment to the device. Furthermore, as the authors in [104] suggested, various approaches could be applied such as unit, black and white boxes, integration, and systems and acceptance testing. Finally, physical tampering mutual authentication could be applied, and encrypted communications and deployment of access control should be developed as the authors in [105] suggested.

There is a clear indication that significant research has been conducted in lightweight authentication schemes for the IoT networks. Studies included in this chapter [93–95] have clearly highlighted the need for more lightweight authentication schemes targeting IoT devices that should not exhaust their hardware capabilities and/or interfere in their functioning. Through extended evaluations, the schemes proposed proved to be a robust against impersonation and MITM attacks. However, there is still room for improvement and extended research. Future directions should be focused on possible lightweight encryption algorithms, digital signatures, and end-to-end encryption to IoT devices so that confidentiality can be strengthen. Furthermore, lightweight blockchain models could be adopted since the exchange of data between different IoT devices is frequent.

Furthermore, regarding authentication, which is an important issue, different authentication schemes could be implemented based on collecting either hardware characteristics of the IoT devices or data from the external environment to authenticate both devices and humans. For device-to-device authentication, hardware and functioning characteristics could be used and for device-to-human authentication, natural body data or human behavior can be tracked to identify the user's identity and validation of his/her usage such as walking pattern and interaction with the device. Through it, even if the device falls into the hands of an adversary, their malicious acts could be quickly detected and potentially prevented. In addition, effective access control could be implemented so even if authentication mechanisms fail, their impact can be potentially reduced due to the authorization constraints effectively in place.

8.8 CONCLUSION

Without a doubt, IoT is going to play a catalytic role in the actual construction of smart cities. IoT networks will continue being the most dominant technology in smart cities offering their services in making the systems more sophisticated and assisting to the city's citizens. However, their individual characteristics, requirements, and technologies would require the deployment of highly intelligent defense mechanisms, otherwise the impact can be catastrophic ending up costing human lives. In this chapter, we discussed how IoT technologies are deployed in smart cities, what benefits and functionality they provide to the smart city and its citizens, and what technologies are used for them to transform the different domains and networks deployed. Then we have explained in detail what vulnerabilities IoTs have and how these can be exploited. Hence, we continue by describing what types of cyberattacks can be conducted and what impact they can have on different smart city networks where IoT technologies are deployed. Furthermore, the most recent and high impact studies on different types of defense mechanisms against such attacks have been presented, always considering the IoT resource constraints. Concluding, future directions were highlighted for new research exploration in the IoT intelligent and lightweight security.

REFERENCES

1. Department of Economic and Social Affairs Population Division, United Nations. World Urbanization Prospects The 2018 Revision. United Nations, 2019. Available: https://population.un.org/wup/Publications/Files/WUP2018-Report.pdf
2. S. P. Mohanty, U. Choppali, and E. Kougianos, "Everything You Wanted to Know About Smart Cities: The Internet of Things is the Backbone," IEEE Consumer Electronics Magazine, vol. 5, no. 3, pp. 60–70, July 2016, doi: 10.1109/MCE.2016.2556879.
3. M. T. Quasim, M. A. Khan, F. Algarni, and M. M. Alshahrani, Fundamentals of Smart Cities. In: Khan M. A., Algarni F., and Quasim M. T. (eds.) Smart Cities: A Data Analytics Perspective. Lecture Notes in Intelligent Transportation and Infrastructure. Springer, Cham, 2021.
4. IBM. IBM Smarter Cities: Creating Opportunities through Leadership and Innovation, White Paper, IBM Corporation, 2014.
5. A. Zanella, N. Bui, A. Castellani, L. Vangelista, and M. Zorzi, "Internet of Things for Smart Cities," IEEE Internet of Things Journal, vol. 1, no. 1, pp. 22–32, Feb. 2014.
6. A. W. Burange and H. D. Misalkar, "Review of Internet of Things in Development of Smart Cities with Data Management & Privacy," 2015 International Conference on Advances in Computer Engineering and Applications, Ghaziabad, 2015, pp. 189–195.
7. E. Theodoridis, G. Mylonas, and I. Chatzigiannakis, "Developing an IoT Smart City framework," IISA 2013, Piraeus, 2013, pp. 1–6.
8. H. Arasteh et al., "IoT-Based Smart Cities: A Survey," 2016 IEEE 16th International Conference on Environment and Electrical Engineering (EEEIC), Florence, 2016, pp. 1–6, doi: 10.1109/EEEIC.2016.7555867.
9. J. P. Lynch and J. L. Kenneth, "A Summary Review of Wireless Sensors and Sensor Networks for Structural Health Monitoring," Shock and Vibration Digest, vol. 38, no. 2, pp. 91–130, 2006.
10. M. A. Khan and K. A. Abuhasel, "Advanced Metameric Dimension Framework for Heterogeneous Industrial Internet of Things," Wiley Computational Intelligence, 2020.
11. T. Alam, M. A. Khan, N. K. Gharaibeh, and M. K. Gharaibeh, Big Data for Smart Cities: A Case Study of NEOM City, Saudi Arabia. In: Khan M. A., Algarni F., and Quasim M. T. (eds.) Smart Cities: A Data Analytics Perspective. Lecture Notes in Intelligent Transportation and Infrastructure. Springer, Cham, 2021.

12. T. Nuortio, J. Kytjoki, H. Niska, and O. Brysy, "Improved Route Planning and Scheduling of Waste Collection and Transport," Expert Systems with Applications, vol. 30, no. 2, pp. 223–232, February 2006.
13. N. Maisonneuve, M. Stevens, M. E. Niessen, P. Hanappe, and L. Steels, "Citizen Noise Pollution Monitoring," In Proceedings of the 10th Annual International Conference on Digital Government Research, Partnerships for Public Innovation. Mexico, 2009, pp. 96–103.
14. S. Lee, D. Yoon, and A. Ghosh, "Intelligent Parking Lot Application Using Wireless Sensor Networks," 2008 International Symposium on Collaborative Technologies and Systems, Chicago, May 1923, 2008, pp. 48–57.
15. W. Kastner, G. Neugschwandtner, S. Soucek, and H. M. Newmann, "Communication Systems for Building Automation and Control," Proceeding of the IEEE, Jun. 2005, vol. 93, no. 6, pp. 1178–1203.
16. M. A. Khan, M. T. Quasim, N. S. Alghamdi, and M. Y. Khan, "A Secure Framework for Authentication and Encryption Using Improved ECC for IoT-Based Medical Sensor Data," IEEE Access, vol. 8, pp. 52018–52027, 2020, doi: 10.1109/ACCESS.2020.2980739
17. J. Mayol, A. Manzoni, F. Calcavecchia, Y. iliev, B. Kabisch, C. Lovis, M. Morgenstern, R. Gomes, G. Gerald, D. Glynos, S. Antonatos, G. Fletcher and P. Jespersen, "Smart Hospitals Security and Resilience for Smart Health Service and Infrastructures," Smart Hospitals About ENISA, 2016, pp. 1–84.
18. C. Lefvy-Benchetton, and E. Darra, "Cyber Security for Smart Cities: An Architecture Model for Public Transport," Cyber Security for Smart Cities, 2015, pp. 1–54.
19. C. Chatzigeorgiou, M. Feidakis, D. G. Kogias, and C. Z. Patrikakis, "Increasing Safety and Security in Public Places Using IoT Devices," 2020 IEEE 6th World Forum on Internet of Things (WF-IoT), New Orleans, LA, USA, 2020, pp. 1–2.
20. M. Madhuri, A. Q. Gill, and H. U. Khan, "IoT-Enabled Smart Child Safety Digital System Architecture," 2020 IEEE 14th International Conference on Semantic Computing (ICSC), San Diego, CA, USA, 2020, pp. 166–169, doi: 10.1109/ICSC.2020.00033.
21. B. Kantarci and H. T. Mouftah, "Trustworthy Sensing for Public Safety in Cloud-Centric Internet of Things," IEEE Internet of Things Journal, vol. 1, no. 4, pp. 360–368, Aug. 2014, doi: 10.1109/JIOT.2014.2337886.
22. M. Z. Saeed, R. R. Ahmed, O. Bin Samin, and N. Ali, "IoT Based Smart Security System Using PIR and Microwave Sensors," 2019 13th International Conference on Mathematics, Actuarial Science, Computer Science and Statistics (MACS), Karachi, Pakistan, 2019, pp. 1–5, doi: 10.1109/MACS48846.2019.9024813.
23. J. Rožman, H. Hagras, J. A. Perez, D. Clarke, B. Müller, and S. F. Data, "Privacy-Preserving Gesture Recognition with Explainable Type-2 Fuzzy Logic Based Systems," 2020 IEEE International Conference on Fuzzy Systems (FUZZ-IEEE), Glasgow, United Kingdom, 2020, pp. 1–8, doi: 10.1109/FUZZ48607.2020.9177768.
24. I. Gao, "Using the Social Network Internet of Things to Mitigate Public Mass Shootings," 2016 IEEE 2nd International Conference on Collaboration and Internet Computing (CIC), Pittsburgh, PA, 2016, pp. 486–489, doi: 10.1109/CIC.2016.073.
25. N. R. Sogi, P. Chatterjee, U. Nethra, and V. Suma, "SMARISA: A Raspberry Pi Based Smart Ring for Women Safety Using IoT," 2018 International Conference on Inventive Research in Computing Applications (ICIRCA), Coimbatore, 2018, pp. 451–454, doi: 10.1109/ICIRCA.2018.8597424.
26. L. Ruiz-Garcia and L. Lunadei, "The Role of RFID in Agriculture: Applications, Limitations and Challenges," Computers and Electronics in Agriculture, vol. 79, no. 1, pp. 42–50, 2011.
27. M. Ghazal, S. Ali, F. Haneefa, and A. Sweleh, "Towards Smart Wearable Real-Time Airport Luggage Tracking," 2016 International Conference on Industrial Informatics and Computer Systems (CIICS), Sharjah, 2016, pp. 1–6, doi: 10.1109/ICCSII.2016.7462422.

28. L. Catarinucci et al., "An IoT-Aware Architecture for Smart Healthcare Systems," IEEE Internet of Things Journal, vol. 2, no. 6, pp. 515–526, Dec. 2015, doi: 10.1109/JIOT.2015.2417684.
29. J. Ahmed, M.Y. Siyal, M. Tayyab, and M. Nawaz, Challenges and Issues in the WSN and RFID. In: RFID-WSN Integrated Architecture for Energy and Delay-Aware Routing. Springer Briefs in Applied Sciences and Technology. Springer, Singapore, 2015.
30. L. Wan, G. Han, L. Shu, N. Feng, C. Zhu, and J. Lloret, "Distributed Parameter Estimation for Mobile Wireless Sensor Network Based on Cloud Computing in Battlefield Surveillance System," IEEE Access, vol. 3, pp. 1729–1739, 2015, doi: 10.1109/ACCESS.2015.2482981.
31. A. Viswanathan, N. B. Sai Shibu, S. N. Rao, and M. V. Ramesh, "Security Challenges in the Integration of IoT with WSN for Smart Grid Applications," 2017 IEEE International Conference on Computational Intelligence and Computing Research (ICCIC), Coimbatore, 2017, pp. 1–4, doi: 10.1109/ICCIC.2017.8524233.
32. L. Zheng, S. Chen, S. Xiang, and Y. Hu, "Research of Architecture and Application of Internet of Things for Smart Grid," 2012 International Conference on Computer Science and Service System, Nanjing, 2012, pp. 938–941, doi: 10.1109/CSSS.2012.238.
33. K. A. Abuhasel and M. A. Khan, "A Secure Industrial Internet of Things (IIoT) Framework for Resource Management in Smart Manufacturing," IEEE Access, vol. 8, pp. 117354–117364, 2020, doi: 10.1109/ACCESS.2020.3004711.
34. M. H. Alharbi and O. H. Alhazmi, "Prototype: User Authentication Scheme for IoT Using NFC," 2019 International Conference on Computer and Information Sciences (ICCIS), Sakaka, Saudi Arabia, 2019, pp. 1–5, doi: 10.1109/ICCISci.2019.8716433.
35. K. Lounis and M. Zulkernine, "Attacks and Defenses in Short-Range Wireless Technologies for IoT," IEEE Access, vol. 8, pp. 88892–88932, 2020, doi: 10.1109/ACCESS.2020.2993553.
36. M. L. Liya and M. Aswathy, "LoRa Technology for Internet of Things (IoT): A Brief Survey," 2020 Fourth International Conference on I-SMAC (IoT in Social, Mobile, Analytics and Cloud) (I-SMAC), Palladam, India, 2020, pp. 8–13, doi: 10.1109/I-SMAC49090.2020.9243449.
37. H. Arasteh, et al. "IoT-Based Smart Cities: A Survey," IEEE 16th International Conference on Environment and Electrical Engineering (EEEIC), Florence, 2016, pp. 1–6, doi: 10.1109/EEEIC.2016.7555867.
38. F. S. Ferraz and C. A. Guimaraes Ferraz, "Smart City Security Issues: Depicting Information Security Issues in the Role of an Urban Environment," 2014 IEEE/ACM 7th International Conference on Utility and Cloud Computing, London, 2014, pp. 842–847, doi: 10.1109/UCC.2014.137.
39. R. Harmon, E. Castro-Leon, and S. Bhide, "Smart Cities and the Internet of Things," Conference: 2015 Portland International Conference on Management of Engineering and Technology (PICMET), 2015, pp. 485–494.
40. A. Imani, A. Keshavarz-Haddad, M. Eslami, and J. Haghighat, "Security Challenges and Attacks in M2M Communications," 2018 9th International Symposium on Telecommunications (IST), Tehran, Iran, 2018, pp. 264–269.
41. M. A. Al Faruque, S. R. Chhetri, A. Canedo, and J. Wan, "Acoustic Side-Channel Attacks on Additive Manufacturing Systems," 2016 ACM/IEEE 7th International Conference on Cyber-Physical Systems (ICCPS), Vienna, 2016, pp. 1–10.
42. S. N. Uke, A. R Mahajan, and R.C Thool, "UML Modeling of Physical and Data Link Layer Security Attacks in WSN," International Journal of Computer Applications, vol. 70, no. 11, May 2013.
43. W. Iqbal, H. Abbas, M. Daneshmand, B. Rauf, and Y. A. Bangash, "An In-Depth Analysis of IoT Security Requirements, Challenges, and Their Countermeasures via Software-Defined Security," IEEE Internet of Things Journal, vol. 7, no. 10, pp. 10250–10276, Oct. 2020, doi: 10.1109/JIOT.2020.2997651.

44. S. Siboni et al., "Security Testbed for Internet-of-Things Devices," IEEE Transactions on Reliability, vol. 68, no. 1, pp. 23–44, March 2019, doi: 10.1109/TR.2018.2864536.
45. A. Syed and R. M. Lourde, "Hardware Security Threats to DSP Applications in an IoT Network," 2016 IEEE International Symposium on Nanoelectronic and Information Systems (iNIS), 2016, pp. 62–66, doi: 10.1109/iNIS.2016.025.
46. S. Koley and P. Ghosal, "Addressing Hardware Security Challenges in Internet of Things: Recent Trends and Possible Solutions," 2015 IEEE 12th International Conference on Ubiquitous Intelligence and Computing (UIC-ATC-ScalCom), 2015, pp. 517–520, doi: 10.1109/UIC-ATC-ScalCom-CBDCom-IoP.2015.105.
47. R. S. Chakraborty, S. Narasimhan, and S. Bhunia, "Hardware Trojan: Threats and Emerging Solutions," 2009 IEEE International High Level Design Validation and Test Workshop, San Francisco, CA, 2009, pp. 166–171, doi: 10.1109/HLDVT.2009.5340158.
48. C. C. Elisan, Malware, Rootkits & Botnets. McGraw Hill, 2013, pp. 9–82.
49. M. Bettayeb, O. A. Waraga, M. A. Talib, Q. Nasir, and O. Einea, "IoT Testbed Security: Smart Socket and Smart Thermostat," 2019 IEEE Conference on Application, Information and Network Security (AINS), Pulau Pinang, Malaysia, 2019, pp. 18–23, doi: 10.1109/AINS47559.2019.8968694.
50. J. Deogirikar and A. Vidhate, "Security Attacks in IoT: A Survey," 2017 International Conference on I-SMAC (IoT in Social, Mobile, Analytics and Cloud) (I-SMAC), Palladam, 2017, pp. 32–37, doi: 10.1109/I-SMAC.2017.8058363.
51. A. Roohi, M. M. Adeel, and M. A. Shah, "DDoS in IoT: A Roadmap Towards Security & Countermeasures," 2019 25th International Conference on Automation and Computing (ICAC), 2019 pp. 1–6.
52. A. Procopiou and N. Komninos, "Current and Future Threats Framework in Smart Grid Domain," 2015 IEEE International Conference on Cyber Technology in Automation, Control, and Intelligent Systems (CYBER), Shenyang, China, 2015, pp. 1852–1857, doi: 10.1109/CYBER.2015.7288228.
53. Z. Chaczko, S. Slehar, and T. Shnoudi, "Game-Theory Based Cognitive Radio Policies for Jamming and Anti-Jamming in the IoT," 2018 12th International Symposium on Medical Information and Communication Technology (ISMICT), Sydney, NSW, Australia, 2018, pp. 1–6, doi: 10.1109/ISMICT.2018.8573725.
54. M. Guizani, A. Gouissem, K. Abualsaud, E. Yaacoub, and T. Khattab, "Combating Jamming Attacks in Multi-channel IoT Networks Using Game Theory," 2020 3rd International Conference on Information and Computer Technologies (ICICT), San Jose, CA, USA, 2020, pp. 469–474, doi: 10.1109/ICICT50521.2020.00081.
55. H. B. Salameh and M. Al-Quraan, "Securing Delay-Sensitive CR-IoT Networking Under Jamming Attacks: Parallel Transmission and Batching Perspective," IEEE Internet of Things Journal, vol. 7, no. 8, pp. 7529–7538, Aug. 2020, doi: 10.1109/JIOT.2020.2985042.
56. X. Tang, P. Ren, and Z. Han, "Jamming Mitigation via Hierarchical Security Game for IoT Communications," IEEE Access, vol. 6, pp. 5766–5779, 2018, doi: 10.1109/ACCESS.2018.2793280.
57. A. Gouissem, K. Abualsaud, E. Yaacoub, T. Khattab, and M. Guizani, "IoT Anti-Jamming Strategy Using Game Theory and Neural Network," 2020 International Wireless Communications and Mobile Computing (IWCMC), Limassol, Cyprus, 2020, pp. 770–776, doi: 10.1109/IWCMC48107.2020.9148376.
58. H. Li, Y. Chen, and Z. He. "The Survey of RFID Attacks and Defenses," 8th International Conference on IEEE Wireless Communications, Networking and Mobile Computing (WiCOM), 2012.
59. H. A. Abdul-Ghani, D. Konstantas, and Md Mahyou, "A Comprehensive IoT Attacks Survey based on a Building-blocked Reference Model," International Journal of Advanced Computer Science and Applications (IJACSA), vol. 9, no. 3, 2018.

60. N. Al-Maslamani and M. Abdallah, "Malicious Node Detection in Wireless Sensor Network using Swarm Intelligence Optimization," 2020 IEEE International Conference on Informatics, IoT, and Enabling Technologies (ICIoT), Doha, Qatar, 2020, pp. 219–224, doi: 10.1109/ICIoT48696.2020.9089527.
61. M. I. Abdullah, M. M. Rahman, and M. C. Roy, "Detecting Sinkhole Attacks in Wireless Sensor Network using Hop Count," International Journal of Computer Network and Information Security, pp. 50–56, 2015.
62. M. M. Iqbal, A. Ahmed, and U. Khadam, "Sinkhole Attack in Multi-sink Paradigm: Detection and Performance Evaluation in RPL based IoT," 2020 International Conference on Computing and Information Technology (ICCIT-1441), Tabuk, Saudi Arabia, 2020, pp. 1–5, doi: 10.1109/ICCIT-144147971.2020.9213797.
63. C. Ioannou and V. Vassiliou, "Accurate Detection of Sinkhole Attacks in IoT Networks Using Local Agents," 2020 Mediterranean Communication and Computer Networking Conference (MedComNet), Arona, Italy, 2020, pp. 1–8, doi: 10.1109/MedComNet49392.2020.9191503.
64. Y. Liu, M. Ma, X. Liu, N. N. Xiong, A. Liu and Y. Zhu, "Design and Analysis of Probing Route to Defense Sink-Hole Attacks for Internet of Things Security," IEEE Transactions on Network Science and Engineering, vol. 7, no. 1, pp. 356–372, 1 Jan.-March 2020, doi: 10.1109/TNSE.2018.2881152.
65. W. Gao et al., "ARP Poisoning Prevention in Internet of Things," 2018 9th International Conference on Information Technology in Medicine and Education (ITME), Hangzhou, China, 2018, pp. 733–736, doi: 10.1109/ITME.2018.00166.
66. S. Ali, M. A. Khan, J. Ahmad, A. W. Malik, and A. ur Rehman, "Detection and Prevention of Black Hole Attacks in IOT & WSN," 2018 Third International Conference on Fog and Mobile Edge Computing (FMEC), Barcelona, Spain, 2018, pp. 217–226, doi: 10.1109/FMEC.2018.8364068.
67. N. Fatima-tuz-Zahra, N. Jhanjhi, S. N. Brohi, and N. A. Malik, "Proposing a Rank and Wormhole Attack Detection Framework using Machine Learning," 2019 13th International Conference on Mathematics, Actuarial Science, Computer Science and Statistics (MACS), Karachi, Pakistan, 2019, pp. 1–9, doi: 10.1109/MACS48846.2019.9024821.
68. M. Tahboush and M. Agoyi, "A Hybrid Wormhole Attack Detection in Mobile Ad-Hoc Network (MANET)," IEEE Access, vol. 9, pp. 11872–11883, 2021, doi: 10.1109/ACCESS.2021.3051491.
69. H. Sinanović and S. Mrdovic, "Analysis of Mirai Malicious Software," 2017 25th International Conference on Software, Telecommunications and Computer Networks (SoftCOM), Split, 2017, pp. 1–5.
70. T. Inoue, K. Hasegawa, M. Yanagisawa, and N. Togawa, "Designing Hardware Trojans and Their Detection Based on a SVM-Based Approach," 2017 IEEE 12th International Conference on ASIC (ASICON), Guiyang, 2017, pp. 811–814, doi: 10.1109/ASICON.2017.8252600.
71. L. Zhang, Y. Dong, J. Wang, C. Xiao, and D. Ding, "A Hardware Trojan Detection Method Based on the Electromagnetic Leakage," China Communications, vol. 16, no. 12, pp. 100–110, Dec. 2019, doi: 10.23919/JCC.2019.12.007.
72. M. Goyal and M. Dutta, "Intrusion Detection of Wormhole Attack in IoT: A Review," 2018 International Conference on Circuits and Systems in Digital Enterprise Technology (ICCSDET), Kottayam, India, 2018, pp. 1–5, doi: 10.1109/ICCSDET.2018.8821160.
73. X. Wang, H. Salmani, M. Tehranipoor, and J. Plusquellic, "Hardware Trojan Detection and Isolation Using Current Integration and Localized Current Analysis," 2008 IEEE International Symposium on Defect and Fault Tolerance of VLSI Systems, Cambridge, MA, USA, 2008, pp. 87–95, doi: 10.1109/DFT.2008.61.
74. Y. Jin and Y. Makris, "Hardware Trojan Detection Using Path Delay Fingerprint," 2008 IEEE International Workshop on Hardware-Oriented Security and Trust, Anaheim, CA, USA, 2008, pp. 51–57, doi: 10.1109/HST.2008.4559049.

75. M. Tehranipoor, H. Salmani, X. Zhang, X. Wang, R. Karri, J. Rajendran, and K. Rosenfeld, "Hardware Trojan Detection Solutions and Design-for-Trust Challenges," IEEE Computer, pp. 64–72, 2011.
76. C. Dong, G. He, X. Liu, Y. Yang, and W. Guo, "A Multi-Layer Hardware Trojan Protection Framework for IoT Chips," IEEE Access, vol. 7, pp. 23628–23639, 2019, doi: 10.1109/ACCESS.2019.2896479.
77. J. Dofe, J. Frey, and Q. Yu, "Hardware Security Assurance in Emerging IoT Applications," 2016 IEEE International Symposium on Circuits and Systems (ISCAS), Montreal, QC, 2016, pp. 2050–2053, doi: 10.1109/ISCAS.2016.7538981.
78. S. J. Johnston, M. Scott, and S. J. Cox, "Recommendations for Securing Internet of Things Devices Using Commodity Hardware," 2016 IEEE 3rd World Forum on Internet of Things (WF-IoT), Reston, VA, 2016, pp. 307–310, doi: 10.1109/WF-IoT.2016.7845410.
79. A. Durand, P. Gremaud, J. Pasquier, and U. Gerber, "Trusted Lightweight Communication for IoT Systems Using Hardware Security," In Proceedings of the 9th International Conference on the Internet of Things (IoT 2019). Association for Computing Machinery, New York, NY, USA, 2019, Article 5, 14. doi: https://doi.org/10.1145/3365871.3365876.
80. A. Kanuparthi, R. Karri, and S. Addepalli, "Hardware and Embedded Security in the Context of Internet of Things," In Proceedings of the 2013 ACM Workshop on Security, Privacy & Dependability for Cyber Vehicles (CyCAR '13). Association for Computing Machinery, New York, NY, USA, 2013, pp. 61–64. doi: https://doi.org/10.1145/2517968.2517976.
81. X. Wang and R. Karri, "Numchecker: Detecting Kernel Control-Flow Modifying Rootkits by Using Hardware Performance Counters," In Proceedings of the 50th ACM/EDAC/IEEE, Design Automation Conference (DAC), 2013, pages 17.
82. D. Champagne and R. B. Lee, "Scalable Architectural Support for Trusted Software," 2010 IEEE 16th International Symposium on High Performance Computer Architecture (HPCA), IEEE, 2010, pp. 1–12.
83. K. Hasegawa, M. Oya, M. Yanagisawa, and N. Togawa, "Hardware Trojans Classification for Gate-Level Netlists Based on Machine Learning," 2016 IEEE 22nd International Symposium on On-Line Testing and Robust System Design (IOLTS), Sant Feliu de Guixols, Spain, 2016, pp. 203–206, doi: 10.1109/IOLTS.2016.7604700.
84. T. Kawamura, M. Fukushi, Y. Hirano, Y. Fujita, and Y. Hamamoto, "An NTP-Based Detection Module for DDoS Attacks on IoT," 2017 IEEE International Conference on Consumer Electronics - Taiwan (ICCE-TW), Taipei, 2017, pp. 15–16, doi: 10.1109/ICCE-China.2017.7990972.
85. P. Shukla, "ML-IDS: A Machine Learning Approach to Detect Wormhole Attacks in Internet of Things," 2017 Intelligent Systems Conference (IntelliSys), London, 2017, pp. 234–240, doi: 10.1109/IntelliSys.2017.8324298.
86. C. Ioannou, V. Vassiliou, and C. Sergiou, "An Intrusion Detection System for Wireless Sensor Networks," 2017 24th International Conference on Telecommunications (ICT), Limassol, 2017, pp. 1–5, doi: 10.1109/ICT.2017.7998271.
87. N. Girnar and S. Kaur, "Intrusion Detection for Adhoc Networks in IOT," 2017 International Conference on Intelligent Computing and Control Systems (ICICCS), Madurai, 2017, pp. 110–114, doi: 10.1109/ICCONS.2017.8250649.
88. L. Almon, M. Riecker, and M. Hollick, "Lightweight Detection of Denial-of-Service Attacks on Wireless Sensor Networks Revisited," 2017 IEEE 42nd Conference on Local Computer Networks (LCN), Singapore, 2017, pp. 444–452, doi: 10.1109/LCN.2017.110.
89. S. Zhao, W. Li, T. Zia, and A. Y. Zomaya, "A Dimension Reduction Model and Classifier for Anomaly-Based Intrusion Detection in Internet of Things," 2017 IEEE 15th International Conference on Dependable, Autonomic and Secure Computing, 15th

International Conference on Pervasive Intelligence and Computing, 3rd International Conference on Big Data Intelligence and Computing and Cyber Science and Technology Congress (DASC/PiCom/DataCom/CyberSciTech), Orlando, FL, 2017, pp. 836–843, doi: 10.1109/DASC-PICom-DataCom-CyberSciTec.2017.141.
90. S. S. Bhunia and M. Gurusamy, "Dynamic Attack Detection and Mitigation in IoT Using SDN," 2017 27th International Telecommunication Networks and Applications Conference (ITNAC), Melbourne, VIC, 2017, pp. 1–6, doi: 10.1109/ATNAC.2017.8215418.
91. H. H. Pajouh, R. Javidan, R. Khayami, A. Dehghantanha, and K. R. Choo, "A Two-Layer Dimension Reduction and Two-Tier Classification Model for Anomaly-Based Intrusion Detection in IoT Backbone Networks," IEEE Transactions on Emerging Topics in Computing, vol. 7, no. 2, pp. 314–323, 1 April–June 2019, doi: 10.1109/TETC.2016.2633228.
92. A. Procopiou, N. Komninos, and C. Douligeris, "ForChaos: Real Time Application DDoS Detection Using Forecasting and Chaos Theory in Smart Home IoT Network," Wireless Communications and Mobile Computing, vol. 2019, Article ID 8469410, 14 pages, 2019. https://doi.org/10.1155/2019/8469410.
93. P. Gope and B. Sikdar, "Lightweight and Privacy-Preserving Two-Factor Authentication Scheme for IoT Devices," IEEE Internet of Things Journal, vol. 6, no. 1, pp. 580–589, Feb. 2019, doi: 10.1109/JIOT.2018.2846299.
94. S. Choi, J. Ko, and J. Kwak, "A Study on IoT Device Authentication Protocol for High Speed and Lightweight," 2019 International Conference on Platform Technology and Service (PlatCon), Jeju, Korea (South), 2019, pp. 1–5, doi: 10.1109/PlatCon.2019.8669418.
95. L. Xiao, Y. Li, G. Han, G. Liu, and W. Zhuang, "PHY-Layer Spoofing Detection With Reinforcement Learning in Wireless Networks," IEEE Transactions on Vehicular Technology, vol. 65, no. 12, pp. 10037–10047, Dec. 2016, doi: 10.1109/TVT.2016.2524258.
96. S. M. Pudukotai Dinakarrao, H. Sayadi, H. M. Makrani, C. Nowzari, S. Rafatirad, and H. Homayoun, "Lightweight Node-level Malware Detection and Network-level Malware Confinement in IoT Networks," 2019 Design, Automation & Test in Europe Conference & Exhibition (DATE), Florence, Italy, 2019, pp. 776–781, doi: 10.23919/DATE.2019.8715057.
97. M. R. Guthaus et al., "MiBench: A Free, Commercially Representative Embedded Benchmark Suite," IEEE 4th Annual International Workshop on Workload Characterization, Austin, Texas, 2001, pp. 3–14.
98. J. L. Henning, "SPEC CPU2006 Benchmark Descriptions," SIGARCH Computer Architecture News, vol. 34, no. 4, pp. 117, Sep 2006.
99. R. Kumar, X. Zhang, W. Wang, R. U. Khan, J. Kumar, and A. Sharif, "A Multimodal Malware Detection Technique for Android IoT Devices Using Various Features," IEEE Access, vol. 7, pp. 64411–64430, 2019, doi: 10.1109/ACCESS.2019.2916886.
100. T. Lei, Z. Qin, Z. Wang, Q. Li, and D. Ye, "EveDroid: Event-Aware Android Malware Detection Against Model Degrading for IoT Devices," IEEE Internet of Things Journal, vol. 6, no. 4, pp. 6668–6680, Aug. 2019, doi: 10.1109/JIOT.2019.2909745.
101. A. Kumar and T. J. Lim, "EDIMA: Early Detection of IoT Malware Network Activity Using Machine Learning Techniques," 2019 IEEE 5th World Forum on Internet of Things (WF-IoT), Limerick, Ireland, 2019, pp. 289–294, doi: 10.1109/WF-IoT.2019.8767194.
102. A. Procopiou and N. Komninos Bio/Nature-Inspired Algorithms in A.I. for Malicious Activity Detection. In: El-Alfy, E-S. M., Elroweissy, M., Fulp, E. W., and Mazurczyk, W. (eds.), Nature-Inspired Cyber Security and Resiliency: Fundamentals, Techniques and Applications. IET, 2019. ISBN 978-1-78561-638-9.

103. J. Su, D. V. Vasconcellos, S. Prasad, D. Sgandurra, Y. Feng, and K. Sakurai, "Lightweight Classification of IoT Malware Based on Image Recognition," 2018 IEEE 42nd Annual Computer Software and Applications Conference (COMPSAC), Tokyo, 2018, pp. 664–669, doi: 10.1109/COMPSAC.2018.10315.
104. K. Sneha and G. M. Malle, "Research on Software Testing Techniques and Software Automation Testing Tools," 2017 International Conference on Energy, Communication, Data Analytics and Soft Computing (ICECDS), Chennai, 2017, pp. 77–81, doi: 10.1109/ICECDS.2017.8389562.
105. M. F. F. Khan and K. Sakamura, "Tamper-Resistant Security for Cyber-Physical Systems with eTRON Architecture," 2015 IEEE International Conference on Data Science and Data Intensive Systems, Sydney, NSW, 2015, pp. 196–203, doi: 10.1109/DSDIS.2015.98.

9 Internet of Vehicles

Design, Architecture, and Security Challenges

Abdullah Alharthi, Qiang Ni, and Richard Jiang
Lancaster University
Lancaster, United Kingdom

CONTENTS

9.1 Introduction 205
9.2 The Architecture of IoV and the Components..206
9.3 Vehicular Ad Hoc Networks...208
    9.3.1 Routing Protocols in VANET...208
        9.3.1.1 Geo-Based Protocols 209
        9.3.1.2 Broadcast-Based Routing Protocols..................................209
        9.3.1.3 Cluster-Based Routing Protocols 209
        9.3.1.4 Geo-Cast Routing Protocols 209
    9.3.2 Topology-Based Routing Protocols 210
        9.3.2.1 Proactive Protocols 210
        9.3.2.2 Reactive Protocol 210
    9.3.3 Hybrid Protocols... 210
9.4 Types of Vehicular Communication 211
    9.4.1 Vehicle-to-Vehicle... 211
    9.4.2 Vehicle-to-Infrastructure 212
    9.4.3 Vehicle-to-Cloud... 212
9.5 Vanet Hardware, Architecture, and Security.. 212
    9.5.1 Hardware Components Within the Vehicular Node
        of an IoV System..212
    9.5.2 Integration of Vehicular Hardware Within the IoV Ecosystem........ 213
    9.5.3 VANET Security Using Blockchain... 214
9.6 Conclusion 216
References.. 217

9.1 INTRODUCTION

The standard of living of people is now improved. Traffic congestions are now more severe as a result of the increment in the number of vehicles. With the uprise of Bluetooth, big data, 5G, cloud computing, sensor technology, and recent tech, there

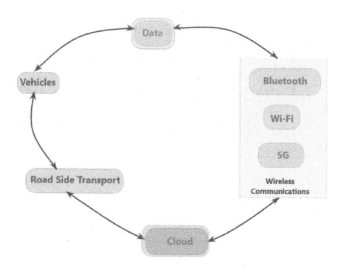

FIGURE 9.1 The basics of the IoV system.

has been significant progress in intelligent transportation. With smart transportation technologies and WLAN's development, IoV has become a very efficient method to alleviate the difficult situation of transportation systems. In the Internet of Vehicles (IoV) system, the environmental information of accompanying road conditions is perceived via the on-board unit (Elkhalil et al., 2020). The data is passed across to the networked vehicles and roadside units (RSUs) via the communication module, as shown in Figure 9.1. The module ensures that vehicle owners have access to road and navigation information. It also reduces traffic congestion possibilities. The IoV's make up the Vehicular Ad-Hoc Networks (VANET) via wireless communication technologies (Shrivastava and Malviya, 2018). The VANET offers several services using instantaneous information shared by cars.

9.2 THE ARCHITECTURE OF IoV AND THE COMPONENTS

The ecosystem of an IoV consists of six major components. They include users (U), vehicle (V), network infrastructure (I), roadside service (R), the person (P), and sensing device (S), as shown in Figure 9.2. Vehicles nearby can establish communication links so relevant information can be exchanged (OICA, 2015). Relevant data includes traffic conditions, physical variables, alerts, etc. (users include persons that access or request a service within the IoV's environment). Personal devices include the devices of individuals (passenger, cyclist, driver, and pedestrian), which are within the IoV's environment (Alam et al., 2015). Network infrastructure includes every network device used in transmitting data. Sensing device involves actuators and sensors used in the gathering of information about the parameters of a vehicle (Vasudev et al., 2020), such as vehicle temperature, fuel consumption, tire pressure, etc. A person's health levels refer to an individual's heart rate, blood pressure, blood oxygen, etc. Environment variables refer to noise level, pollution, weather conditions, etc.

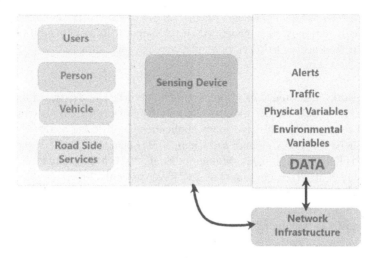

FIGURE 9.2 Architecture and components of IoV.

Roadside devices refer to infrastructures in the transportation sector like information screens, traffic lights, or radars with the capability to pass vital information about accidents, traffic jams, or possible detours across.

Devices and connected devices are essential concepts of the IoV concept. It is a mobile system that permits the exchange of information via Vehicle to Vehicle (V2V), Vehicle and Person (V&P), Vehicle and Roadside (V&R), Device to Device (D2D), and Vehicle and Device (V&D), as shown in Figure 9.3. (Nanjie, 2011). With the IoV, sturdy interactions are promoted between vehicles and humans. It enhances human abilities like hearing, spatial awareness, vision, and sensitivity. It also improves safety, vehicle intelligence, experience, and energy. It impacts several markets like the consumer vehicle market, consumer lifestyle, and consumer behavior (APEC, 2014; McKinsey and Company, 2013).

Implementing the IoV requires devices to interact and communicate with other devices in the infrastructure using several technologies dependent on the type of device, network, etc. Devices include tablets, personal communication devices, and sensors. Network type includes Wireless Sensor Network (WSN), Personal Area

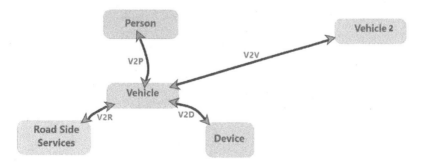

FIGURE 9.3 Exchange mechanisms within IOVs.

Network (PAN), etc. This presents interaction scenarios that are complex whereby several challenges must be attended to. The challenges are incompatibility between devices, limited access to services of that process, and storage data (Wang et al., 2020). As such, integrating IoV with:

- Networks in existence like Long-Term Evolution (LTE), Wi-Fi, cellular, etc.
- Emerging technologies like the Internet of Things (IoT), cloud computing, and connected vehicles has many applications in different fields (Abuhasel and Khan, 2020; Alghamdi and Khan, 2021; Khan et al., 2020; Khan et al., 2021; Khalifa et al., 2021; Mukherjee et al., 2021; Munusamy et al., 2021; Nandy et al., 2021; Quasim et al., 2021; Verma et al., 2021)
- Infrastructure related to transportation like signs, RSUs, and traffic lights.
- Seamless integration is essential to ensure effective communication within the IoV's environment.

9.3 VEHICULAR AD HOC NETWORKS

VANET originates from the "Internet of Things." The IoT is the major technology for driving autonomously. The IoV system integrates inter-vehicular communication and improves communication technology sites (Rana et al., 2014). The three main communication components of the VANET system are vehicular mobile internet, intra-vehicular communication, and inter-vehicular communication (Priyan and Devi, 2019).

9.3.1 Routing Protocols in VANET

VANET is highly mobile and fluid, hence, it is necessary to use the right routing protocols. The network packets are sent from one vehicle to another that moves with speed (Saini and Singh, 2016). Routing protocols are playing an important role for the communications among peer vehicles in VANET. The classification of routing protocols in VANET is shown in Figure 9.4. The vehicle's density increases and

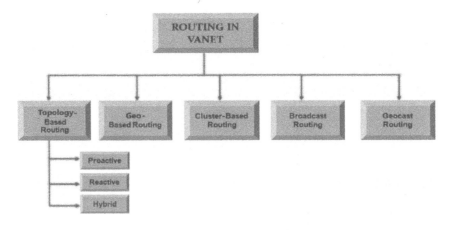

FIGURE 9.4 Routing protocols in VANET.

Internet of Vehicles

decreases so, several difficulties are linked to routing protocols (Ahmed et al., 2017). Routing protocols are considered into the following.

9.3.1.1 Geo-Based Protocols

Here, communication will be done by a source via network address and geographical destinations. Load Balancing Routing Protocol (LBRP) is used to configure and calculate routes based on information and node location. So, building route tables is needless. The components involved in this protocol are forwarding services, beaconing, and site (Contreras-Castillo et al., 2018). The disadvantage of this geo-based protocol is that it needs Global Positioning System (GPS) to ascertain vehicles' location (Rana et al., 2014). Satellite signals are weak whenever cars go into places like tunnels; however, they work perfectly in open highway environments. This protocol has its advantages too. It works efficiently in high mobility environs. The protocols include Distance Routing Effect Algorithm for Mobility (DREAM) and Greedy Perimeter Stateless Routing.

9.3.1.2 Broadcast-Based Routing Protocols

The broadcast-based routing protocols floods the data packet to every part of the VANET to every node available in the broadcast domain. Whenever the destination host is not within the range, that is when this protocol is used. These protocols are mostly utilized with safety-inclined applications like emergency and weather condition warning messages, etc. (Hu et al., 2017). Examples of this protocol are Position Aware Reliable Broadcasting Protocol (POCA), Distributed Vehicular Broadcast Protocol (DV-CAST), and Density Aware Reliable Broadcasting Protocol (DECA). An advantage of this protocol is that it is reliable. The downside is that several identical packets get to the nodes, and the protocol system takes considerable bandwidth.

9.3.1.3 Cluster-Based Routing Protocols

In the cluster-based routing protocol, vehicles with similar characteristics like direction and speed are joined in a single closer. If there should be communication within nodes in the cluster, it must be communicated locally through a direct path (Mohammed et al., 2014). When the vehicle node needs to communicate with an external node, it is assisted by the cluster head to reach its destination. Because of how scalable the factors are, network providers choose it. A disadvantage of this protocol is the delay in traffic that causes feedback. An example of this protocol is Clustering for Open Inter Vehicular Communication Network (COIN).

9.3.1.4 Geo-Cast Routing Protocols

This involves two significant zones called Zone of Relevance (ZOR) and Zone Of Forwarding (ZOF). ZOR is an area, especially for the nodes of that environment (Song and Shoaib, 2014). The primary objective of geo-cast routing protocols is to ensure effective communication between vehicles available in ZOR (Paul, 2012). If the vehicle absent in ZOR wishes to communicate with the source vehicle, it will belong to ZOF. Every ZOF vehicle must share information with other ZORs. As a result of time zone, there can be regular disconnection that could cause drawbacks.

9.3.2 Topology-Based Routing Protocols

9.3.2.1 Proactive Protocols

Here, the routing table is updated frequently because the regular protocols are computed. The algorithm used is the Bellman-Ford Algorithm. In this algorithm, every node keeps the information of the other (Kumari, 2017). An advantage is knowing the routes whenever data is to be sent by a packet. An example of this protocol is Optimized Link-State Routing (OLSR) and Destination Sequenced Distance Vector (DSDV) protocol.

OLSR is an advanced type of link-state routing (LSR). LSR works in such a manner that any alteration in the topology will be shared to all network nodes to heighten network overhead. OSLR controls messages like "hello" and handles topology messages. Statements like "hello" are used to source out information about the connection status. In contrast, messages to manipulate topology share neighbor-to-neighbor information using the multi-point relay (MPR) selected list. As a result of using MPR, overload had decreased like when it was a simple LSR. DSDV is an advanced form of the Bellman-Ford Algorithm. This algorithm is used to test the issue of routing looping by maintaining the sequence number details of every node.

9.3.2.2 Reactive Protocol

The reactive protocol does not have information regarding all nodes (Zhang and Sun, 2016). The protocol stores information about nodes in the protocol. Examples include Dynamic Source Control Routing (DSR), Ad-hoc On-Demand Distance Vector Reactive protocols (AODV), and Dynamic MANET on Demand (DYMO) protocol. AODV is used mostly in VANET. It contains the next-hop for destination nodes. Every routing stable stays for a given period. If no routes are demanded within a specific timeframe, routes will fade away, and new ones will be made available when data is to be received from destination nodes to source nodes. The routes to be confirmed come from its routing table. If the routing information is available on the routing table, packets will be successfully transferred to their destinations. If the routing information is unavailable, route discovery requests will be broadcasted to neighbors. This mobile ad-hoc protocol is appropriate in high mobility traffic. AODV reduces broadcasting overhead. The process of discovering routes is done on demand (Sallam and Mahmoud, 2015).

DSR is a very productive routing protocol. Primarily, it is made for Wireless Ad-Hoc Network (WANET). It is a protocol that organizes and configures itself without supervision. The primary functions are route maintenance and discovery. After AODV, the DYMO protocol was designed. The protocol can be used as a proactive and reactive protocol (Spaho et al., 2013). Moreover, the methodology of discovering routes is demanded whenever it is required.

9.3.3 Hybrid Protocols

This is the merging of reactive and proactive routing protocols to reduce delays and overhead that result from broadcasting topology information. Using this hybrid protocol, there is an improvement in network scalability and efficiency. The disadvantage is that there is a rise in latency for a new route navigation. A typical example of a hybrid protocol is Zone Routing Protocol (ZRP).

9.4 TYPES OF VEHICULAR COMMUNICATION

A major trend in the environment of IoV is the addressing of communication mishaps among different devices in various domains like safety, traffic control, and infotainment. Communication applications are limited in their interoperability due to issues concerning availability, privacy, and accessibility. This is why the applications operate independently. So many attempts have been made to increase interoperability and reduce application silos (Bonomi, 2013; Nanjie, 2011; Wan et al., 2014). The various attempts pay attention to developing multiple interaction platform frameworks that enable various devices and vehicles to communicate on the IoV environments (Xu et al., 2021). The major types of vehicular communication include V2V, vehicle-to-infrastructure (V2I), and vehicle-to-cloud (V2C), as shown in Figure 9.5 (Pothirasan et al., 2020).

9.4.1 Vehicle-to-Vehicle

This had to do with exchanging information wirelessly about the position and speed of surrounding vehicles to avoid crashes, ease congestion, and improve the environment (Saxena, 2019). Every node will send, retransmit signals, and a mesh network, meaning every node (car, smart traffic signal, etc.) could send, capture, and retransmit signals (Howard, 2014), which is essential for developing countries (Zorkany, 2020).

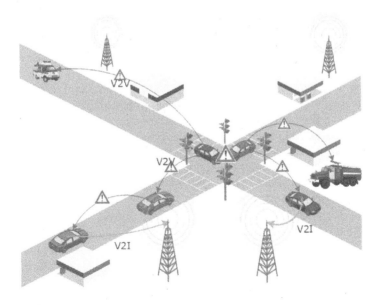

FIGURE 9.5 Vehicular communications.

9.4.2 Vehicle-to-Infrastructure

V2I is a data exchange between road infrastructure and vehicles (Baldessari et al., 2006), with infrastructures consisting of signage, parking meters, lane makers, RFID readers, cameras, and streetlights. It is an Intelligent Transportation System's (ITS) next-generation (Intelligent Transportation Systems – Vehicle to Infrastructure (V2I) Deployment Guidance and Resources, n.d.).

9.4.3 Vehicle-to-Cloud

V2C supports security based on cloud, entertainment, connected car serviced, and personal information. It handles everything ranging from security authentication to firmware updates to securing reservations and locating parking spots (Mukherjee et al., 2021). V2C has the potential to handle the vulnerable futuristic safety of road.

9.5 VANET HARDWARE, ARCHITECTURE, AND SECURITY

The system's hardware implementation is carried out through the nRF and Arduino board modules. For easy comprehension, multiple nodes of the system are formed to copy a real-life scenario. VANET's hardware implementation is performed using microcontrollers, NRF24L01 modules, and other sensors. The components can be used to create an anti-collision system functioning as a precrash and forward warning system. It works to prevent the severity of a collision. The components that build this system are Arduino Uno boards, LCD module, GPS module, IR sensor, NRF24L01 module, and buzzer.

9.5.1 Hardware Components Within the Vehicular Node of an IoV System

Every computerized system needs a motherboard, which enables the entire system to function properly (Duncan, 2017). It consists of integrated circuits (ICs) and communication nodes processing signals. According to Mlakic et al. (2019), Arduino devices are used to build motherboards. Although Arduino devices have been largely used in creating prototypes, they have also been used in production environments (Chen et al., 2018), and can be used in creating sensing devices in an IoV system. They consist of a microcontroller board with a processor named ATmega328P, as shown in Figure 9.6. There are 14 digital IO pins (six are used as PWM outputs and the remaining six as analog inputs), a USB connection, reset button, 16 MHz quartz crystal, power jack, and ICSP header (Belwal and Rakesh, 2019). Arduino boards and nRF modules can be used in the creation of the network.

The NRF2L01 module is a device that carries out several transmitting and receiving functions. It can communicate directly with about six different modules through routing (Belwal and Rakesh, 2019). This component can be used in communicating with various nodes by creating a common network between nodes. It is a transceiver that enables wireless communication within Arduino boards (Guerrero-Ibañez et al., 2011). The GPS module is used in the location of vehicles using satellite positions (Keerthika

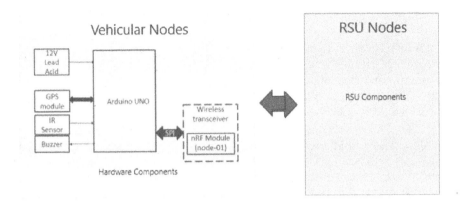

FIGURE 9.6 The IoV system and hardware components within the vehicular nodes.

et al., 2017). The data is processed using Arduino. The coordinates that consist of axis x and y get transmitted to the RSUs. The GPS module works with the NRF2L01 module to create a relative position of the vehicle at every instant. Finally, each intelligent vehicle has onboard unit and some other sensors, as shown in Figure 9.7.

9.5.2 Integration of Vehicular Hardware Within the IoV Ecosystem

The hardware components are a part of the vehicle nodes. An integration exists between the number of RSUs and vehicle nodes within the IoV system. A wireless network consisting of two nodes is formed with the RSU and vehicular nodes (Liang et al., 2015). Within the vehicular nodes, a GPS component is linked to vehicles,

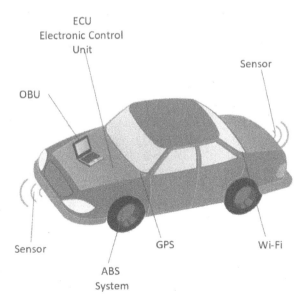

FIGURE 9.7 Main components of vehicle in VANET.

which takes notes of the GPS information. The data is moved to the RSU close to it (Belwal and Rakesh, 2019). The LCD component is employed in showing messages sent from other nodes to this node.

A vehicle gets connected to a network when it gets within range of an RSU. The destination and stats of vehicles are shared with RSUs. In turn, the RSUs reply with routes that can be followed and the route's status. The Shortest Path Routing Algorithm (SPRA) is used to discover the shortest paths that can be used in communicating data (Belwal and Rakesh, 2019). The devices to be used in computing are placed apart from each other. The system is triggered when the vehicle gives a destination. Then, the shortest path is computed by both the vehicle's range and the RSUs. Euclidean distance is the foundation of the algorithm employed in discovering the shortest route from source to destination (Lavor et al., 2011). The path can be used in sharing information like accident occurrence and traffic status in the route.

The RSUs are responsible for the monitoring of coordinates and networks. Every vehicle in the IoV is within the range of the RSU (Sherazi et al., 2019). So, the roadside node acts as a server node. Vehicular nodes are sometimes referred to as client nodes. In the client nodes, events are performed based on written code(s). The server node plays the role of the system's RSU. It monitors the network and also sends important messages to vehicles that go into the network.

9.5.3 VANET Security Using Blockchain

The IoV is a complicated system comprising network facilities like people and vehicles. Communication is done via a wireless network, so security issues pose a significant threat to the system (Alaya, 2020). Therefore, it is a matter of great urgency to provide services and protect its users' privacy. Vehicle drivers can consider the distance. Drivers can place routes in line with the current traffic environment improving the general traffic system efficiency. As a result of IoV's service particularity, it is a necessity for vehicles to communicate and share location information often. There are several security risks involved. Vehicles' shared data can be analyzed and collected by attackers (Rupareliya et al., 2016). Privacy can be stored, and owners of vehicles can be directionally tracked via driving trajectory. A system for protecting privacy based on blockchain has been proposed for the IoV. Blockchain is combined with IoV by developing an efficient "key agreement algorithm" and "two-way authentication" (Jiang et al., 2019). This is done via signature algorithm and encryption. So, IoV's old-fashioned dependency problem is solved.

To resolve the matter of communication and security efficiency, the privacy protection system using blockchain is proposed, which resolves the central failure problem that arises as a result of excessive access. This excessive access can occur through a distributed mechanism (where public key of vehicle uses for information management). The blockchain is characterized as tamper-proof, so it can stop a vehicle's general key details stored in the chain from being messed with. The system uses a voting system with malicious behavior in identifying messages sent by vehicles to the chain's node (Kim, 2019). This curtails the vehicle from publishing illegal location details when an attacker has malicious intent.

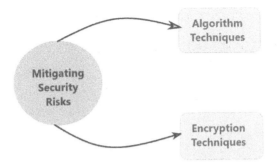

FIGURE 9.8 Mitigating the security risk through algorithm and encryption techniques with blockchain.

Essentially, blockchain is a distributed and decentralized database technology. Blockchain preserves a data block that is structurally chained among participating nodes. It grows continuously and immutably records data based on cryptography (Li et al., 2020). Every data block can be divided into block body and blockhead. In the blockhead, each block is linked with hash values. Each block transaction has an association with the Merkle root. Blockchain will analyze the transaction information to every detail via the agreeing mechanism (Bartoletti and Pompianu, 2017). A decentralized storage process is formed when clients store recent transaction information. This way, the system will not be affected when nodes disappoint. Figure 9.8 shows two methods to mitigate the security risk in communications among peer vehicles. The system uses cryptographic algorithms and blockchain to design effective exchange strategies for keys and two-way authentications so that communication costs can be reduced. Simultaneously, blockchain can be used by the system's vehicles to communicate anonymously with the providers of service locations. It will prevent service providers from stealing the privacy of the vehicles. The IoV privacy protection based on blockchain is cheaper, easier to maintain, and more secure than the conventional IoV scheme (Abbad and Godse, 2016).

As shown in Figure 9.9, IoV's architecture system is based on blockchain. It is crucial that any information obtained from real-time traffic, like safety messages, is processed in the IoV. As a result of the vehicle's restricted computing power, RSURSUs are used by the system as a host for the Blockchain network. The RSUs have forwarding, routing, and several beneficial computing power functions (Kaur and Dhaliwal, 2016). The RSUs forward received data to its server by adopting a cloud computing method with sufficient storage space and strong real-time computing power. Sufficient storage capacity provides the blockchain system with the necessary computing power. The cloud server ensures RSU has a fast database interface for receiving information. Cloud computing assures data analysis and storage in the IoV system (Lopez, Siller and Huerta, 2017).

Blockchain is used in protecting users' private information so it cannot be altered illegally. Using the blockchain system, the significant nodes can be held accountable if any problem were to arise. Law Enforcement Agency (LEA) is also included in the blockchain system, so user information and information generated from the

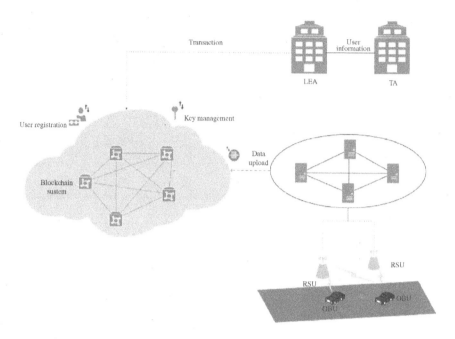

FIGURE 9.9 Vehicular communication secured with blockchain. (From Lopez, Siller and Huerta, 2017.)

technical analysis can be reviewed. The info obtained is signed and then forwarded to the blockchain network.

9.6 CONCLUSION

The IoV system consists of layers working together to provide essential services and achieve a common goal. The first layer has to do with vehicles, communication links, and V2V communication software with network protocols. Infrastructure is the second layer that defines tech that ensures connectivity among participants. The operation layer becomes relevant as it ensures and verifies compliance with policies that should be applied to monitor information flow and management. The cloud layer provides the platform to transact services like data, voice, and enterprise video on demand. The connection layer ensures interoperability, and every network support communication model such as V2V, V2I, V2R, and V2P. It sends information across to the cloud layer and provides the needed computational power necessary in satisfying the requirements needed by vehicles like transferring communication spectrum, searching information, analyzing information, and storing data.

The IoV basic architecture includes client, vehicular, location, cloud, and connection within an ITS. In the client layer, every sensor within and without the vehicle has the responsibility of measuring vehicle position, speed, tire and oil pressure, noise, pollution levels, forward obstacle, proximity, etc. All these are responsible for capturing information about events that happen in the vehicle. This includes intra-vehicular, driving patterns, and inter-vehicular connections with other vehicles.

The vehicular layer manages the internal sensors of vehicles. It processes information such as psychological and ambient parameters, e.g., the vehicle's passengers, via short-range wireless technology. The architecture allows information to be exchanged with nearby vehicles. If vehicles are not within range, exchanging information is done through multi-hop communications while being supported by RSUs. Services are hosted by the cloud layer that gives access to previous traffic information. This layer enables the load to be balanced across several interconnected cloud systems.

REFERENCES

Abbad, S. A. & Godse, P. (2016). Secure and privacy preserving navigation in VANET. *International Journal of Science and Research (IJSR)*, 5(3), 997–1001. https://doi.org/10.21275/v5i3.nov162079

Alam, K. M., Saini, M., & Saddik, A. E. (2015). Toward social internet of vehicles: Concept, architecture, and applications. *IEEE Access*, 3, 343–357. https://doi.org/10.1109/access.2015.2416657

Abuhasel, K. A. & Khan, M. A. (2020). A secure industrial internet of things (IIOT) framework for resource management in smart manufacturing, *IEEE Access*, 8, 117354–117364. doi: 10.1109/ACCESS.2020.3004711.

Alghamdi, N. S. & Khan, M. A. (2021). Energy-efficient and blockchain-enabled model for internet of things (IoT) in smart cities, *Computers, Materials & Continua*, 66, 3, 2509–2524.

Alaya, B. (2020). Efficient privacy-preservation scheme for securing urban P2P VANET networks. *Egyptian Informatics Journal*, e3. https://doi.org/10.1016/j.eij.2020.12.002

APEC. (2014). White paper of Internet of Vehicles. 50th Telecommunications and information working group meeting, Brisbane, Australia

Ahmed, S., Ur Rehman, M., Ishtiaq, A., Khan, A., Ali, A., & Begum,S. (2017). Attack resistant VANET data communication using vehicle movement behavior analysis. *International Journal of Modern Trends in Engineering & Research*, 4(9), 100–106. https://doi.org/10.21884/ijmter.2017.4285.jusvx

Baldessari, R., Andreas, F., Alfredo, M., Justino, S., & Rui, A. (2006). Flexible connectivity management in vehicular communication networks. *Proceedings of 3rd International Workshop on Intelligent Transportation (WIT)*. pp. 211–216, Hamburg, Germany.

Bartoletti, M. & Pompianu, L. (2017). An empirical analysis of smart contracts: Platforms, applications, and design patterns, in *Financial Cryptography and Data Security*, M. Brenner, Ed., vol. 10323 of Lecture Notes in Computer Science, Springer, Cham, Switzerland.

Belwal, M. & Rakesh, G. (2019). Hardware implementation of VANET communication based collision warning system. *Proceedings of the 4th International Conference on Communication and Electronics Systems ed.* IEEE.

Bonomi, F. (2013). *The Smart and Connected Vehicle and the Internet of Things*. San José, CA: WSTS.

Chen, L., Zhang, J., & Wang, Y. (2018). Wireless car control system based on ARDUINO UNO R3. *2018 2nd IEEE Advanced Information Management, Communicates, Electronic and Automation Control Conference (IMCEC)*. https://doi.org/10.1109/imcec.2018.8469286

Contreras-Castillo, J., Zeadally, S., & Guerrero-Ibanez, J. A. (2018). Internet of Vehicles: Architecture, protocols, and security. *IEEE Internet of Things Journal*, 5(5), 3701–3709. https://doi.org/10.1109/jiot.2017.2690902

Duncan, W. D. (2017). Ontological distinctions between hardware and software. *Applied Ontology*, *12*(1), 5–32. https://doi.org/10.3233/ao-170175

Elkhalil, A., Zhang, J., Elhabob, R., & Eltayieb, N. (2020). An efficient signcryption of heterogeneous systems for Internet of Vehicles. *Journal of Systems Architecture*, 101885. https://doi.org/10.1016/j.sysarc.2020.101885

Guerrero-Ibañez, A., Contreras-Castillo, J., Barba, A., & Reyes, A. (2011). A QoS-based dynamic pricing approach for services provisioning in heterogeneous wireless access networks. *Pervasive and Mobile Computing*, *7*(5), 569–583. doi: 10.1016/j.pmcj.2010.10.003.

Keerthika, C., Singh, M., & Tamizharasi, T. (2017). Tracking system for vehicles using GPS, GSM and GPRS. *Research Journal of Engineering and Technology*, *8*(4), 453. https://doi.org/10.5958/2321-581x.2017.00078.2.

Khan, M. A., Quasim, M. T., Alghamdi, N. S., & Khan, M. Y. (2020) A secure framework for authentication and encryption using improved ECC for IoT-based medical sensor data," *IEEE Access*, *8*, 52018–52027. doi: 10.1109/ACCESS.2020.2980739.

Khan, W, U, Li., X., Ihsan, A., Khan, M. A., Menon, V. G. &Ahmed, M. (2021). NOMA-enabled optimization framework for next-generation small-cell IoV networks under imperfect SIC decoding, *IEEE Transactions on Intelligent Transportation Systems*. doi: 10.1109/TITS.2021.3091402.

Khalifa, M., Algarni, F., Khan, M. A., Ullah, A., & Aloufi, K. (2021). A lightweight cryptography (LWC) framework to secure memory heap in Internet of Things. *Alexandria Engineering Journal*, *60*(1), 1489–1497, ISSN 1110-0168. https://doi.org/10.1016/j.aej.2020.11.003.

Howard, B. (2014, February 6). V2V: What are vehicle-to-vehicle communications and how do they work? ExtremeTech. https://www.extremetech.com/extreme/176093-v2v-what-are-vehicle-to-vehicle-communications-and-how-does-it-work

Hu, S., Jia, Y., & She, C. (2017). Performance analysis of VANET routing protocols and implementation of a VANET terminal. *International Conference on Computer Technology, Electronics and Communication (ICCTEC)*, *2017*, 1248–1252. doi: 10.1109/ICCTEC.2017.00272.

Intelligent Transportation Systems – Vehicle to Infrastructure (V2I) Deployment Guidance and Resources. (n.d.). US DOT. https://www.its.dot.gov/v2i/

Jiang, T., Fang, H., & Wang, H. (2019). Blockchain-based internet of vehicles: Distributed network architecture and performance analysis. *IEEE Internet of Things Journal*, *6*(3), 4640–4649. https://doi.org/10.1109/jiot.2018.2874398

Kaur, P. & Dhaliwal, D. K. (2016). Various approaches of VANET routing and attack detection. *International Journal of Science and Research (IJSR)*, *5*(2), 1999–2002. https://doi.org/10.21275/v5i2.nov161429

Kim, S. (2019). Impacts of mobility on performance of blockchain in VANET. *IEEE Access*, *7*, 68646–68655. https://doi.org/10.1109/access.2019.2918411

Kumari, S. (2017). Simulation of VANET routing using A-star algorithm. *International Journal of Trend in Scientific Research and Development*, *1*(4), 465–469. https://doi.org/10.31142/ijtsrd178

Lavor, C., Liberti, L., Maculan, N., & Mucherino, A. (2011). The discretizable molecular distance geometry problem. *Computational Optimization and Applications*, *52*(1), 115–146. https://doi.org/10.1007/s10589-011-9402-6

Li, X., Jiang, P., Chen, T., Luo, X., & Wen, Q. (2020). A survey on the security of blockchain systems. *Future Generation Computer Systems*, *107*, 841–853.

Liang, W., Li, Z., Zhang, H., Wang, S., & Bie, R. (2015). Vehicular ad hoc networks: Architectures, research issues, methodologies, challenges, and trends. *International Journal of Distributed Sensor Networks*, *11*(8), 745303. https://doi.org/10.1155/2015/745303

Lopez, H. J. D., Siller, M., & Huerta, I. (2017). Internet of vehicles: Cloud and fog computing approaches. *IEEE International Conference on Service Operations and Logistics, and Informatics (SOLI)*. 211–216. doi: 10.1109/SOLI.2017.8120996.

McKinsey & Company. (2013). The road to 2020 and beyond – What's driving the global automotive industry. McKinsey & Company. http://www.mckinsey.com/client_service/automotive_and_assembly/latest_thinking

Mlakic, D., Nikolovski, S., & Baghaee, H. R. (2019). An open-source hardware/software IED based on IoT and IEC 61850 Standard. *2019 2nd International Colloquium on Smart Grid Metrology (SMAGRIMET)*, e33. https://doi.org/10.23919/smagrimet.2019.8720361

Mohammed, F., Mohamed, O., Abedelhalim, H., & Abdellah, E. (2014). Efficiency evaluation of routing protocols for Vanet. *2014 Third IEEE International Colloquium in Information Science and Technology (CIST)*. 410–414. doi: 10.1109/CIST.2014.7016655.

Mukherjee, A., Goswami, P., Khan, M. A., Manman, L., Yang, L., & Pillai, P. (2021). Energy-efficient resource allocation strategy in massive IoT for industrial 6G applications. *IEEE Internet of Things Journal*, 8(7), 5194–5201. doi: 10.1109/JIOT.2020.3035608.

Munusamy, A. et al. (2021). Service deployment strategy for predictive analysis of FinTech IoT applications in edge networks. *IEEE Internet of Things Journal*. doi: 10.1109/JIOT.2021.3078148.

Nandy, S., Adhikari, M., Khan, M. A., Menon, V. G., & Verma, S., (2021). An intrusion detection mechanism for secured IoMT framework based on Swarm-Neural Network. *IEEE Journal of Biomedical and Health Informatics*. doi: 10.1109/JBHI.2021.3101686.

Nanjie, L. (2011). Internet of Vehicles your next connection. *WinWin Magazine*, Issue 11, HUAWEI.

OICA. (2015). Number of passenger cars and commercial vehicles in use worldwide from 2006 to 2014. Organisation Internationale des Constructeurs d'Automobile (OICA). http://www.statista.com/281134/number-of-vehicles-in-use-worldwide/

Quasim, M.T., Alkhammash, E.H., Khan, M.A. et al. (2021). Emotion-based music recommendation and classification using machine learning with IoT Framework. *Soft Computing*, 25, 12249–12260. https://doi.org/10.1007/s00500-021-05898-9

Paul, Bijan. (2012). Survey Over VANET Routing Protocols For Vehicle To Vehicle Communication. IOSR Journal Of Computer Engineering, 7(5), 1–9. IOSR Journals. https://doi.org/10.9790/0661-0750109.

Pothirasan, N., Pallikonda Rajasekaran, M., & Muneeswaran, V. (2020). Priority Based an Accident Prevention System Using V2V Communication. *Journal of Advanced Research in Dynamical and Control Systems*, 51(SP3), 480–485. https://doi.org/10.5373/jardcs/v12sp3/20201282

Priyan, M. & Devi, G. (2019). A survey on Internet of vehicles: Applications, technologies, challenges and opportunities. *International Journal of Advanced Intelligence Paradigms*, 12(½), 98–119.

Rana, S., Rana, S., & Purohit, K. C. (2014). A review of various routing protocols in VANET. *International Journal of Computer Applications*, 96(18), 28–35. https://doi.org/10.5120/16896-6946

Rupareliya, J., Vithlani, S., & Gohel, C. (2016). Securing VANET by preventing attacker node using watchdog and Bayesian network theory. *Procedia Computer Science*, 79, 649–656. https://doi.org/10.1016/j.procs.2016.03.082

Saini, M. & Singh, H. (2016). VANET, its characteristics, attacks and routing techniques: A survey. *International Journal of Science and Research (IJSR)*, 5(5), 1595–1599. https://doi.org/10.21275/v5i5.nov163726

Sallam, G. & Mahmoud, A. (2015). Performance evaluation of OLSR and AODV in VANET cloud computing using fading model with SUMO and NS3, *Proceedings of International Conference on Cloud Computing (ICCC)*. pp. 1–5.

Saxena, A. (2019, December 11). *Vehicle-to-Vehicle Communication: Let the car message while driving, not you!* Crazyblog. https://www.einfochips.com/blog/vehicle-to-vehicle-communication-let-the-car-message-while-driving-not-you/

Sherazi, H. H. R., Khan, Z. A., Iqbal, R., Rizwan, S., Imran, M. A., & Awan, K. (2019). A heterogeneous IoV architecture for data forwarding in vehicle to infrastructure communication. *Mobile Information Systems, 2019*, 1–12. https://doi.org/10.1155/2019/3101276

Shrivastava, V. & Malviya, S. (2018). A secure multi agent VANET model to improve communication services in VANET. *International Journal of Trend in Scientific Research and Development, 2*(3), 2562–2566. https://doi.org/10.31142/ijtsrd12833

Song, Wang-Cheol & Shoaib, Muhammad. (2014). A ZHLS based geocasting for broadcasting messages in VANET. *International Journal Of Computer And Electrical Engineering, 6*, 4, 333–336. https://doi.org/10.7763/ijcee.2014.v6.848.

Spaho, E., Ikeda, M., Barolli, F., Xhafa, F., Younas, M., & Takizawa, M. (2013). Performance evaluation of OLSR and AODV protocols in a VANET, *Proceedings of IEEE 27th International Conference on Advanced Information Networking and Applications (AINA).* pp. 577–582.

Vasudev, H., Deshpande, V., Das, D., & Das, S. K. (2020). A lightweight mutual authentication protocol for V2V communication in Internet of Vehicles. *IEEE Transactions on Vehicular Technology, 69*(6), 6709–6717. https://doi.org/10.1109/tvt.2020.2986585

Vehicle-to-Vehicle Communication. (2019, December 18). NHTSA. https://www.nhtsa.gov/technology-innovation/vehicle-vehicle-communication

Verma, S., Kaur, S., Khan, M. A., & Sehdev, P. S. (2021). Toward green communication in 6G-enabled massive Internet of Things. *IEEE Internet of Things Journal, 8*(7), 5408–5415. doi: 10.1109/JIOT.2020.3038804.

Wan, J., Zhang, D., Zhao, S., Yang, L., & Lloret, J. (2014). Context-aware vehicular cyber-physical systems with cloud support: Architecture, challenges, and solutions. *IEEE Communications Magazine, 52*(8), 106–113. doi: 10.1109/MCOM.2014.6871677.

Wang, X., Han, S., Yang, L., Yao, T., & Li, L. (2020). Parallel Internet of Vehicles: ACP-based system architecture and behavioral modeling. *IEEE Internet of Things Journal, 7*(5), 3735–3746. https://doi.org/10.1109/jiot.2020.2969693.

Xu, L., Zhou, X., Khan, M. A., Li, X., Menon, V. G., & Yu, X. (2021). Communication quality prediction for Internet of Vehicle (IoV) networks: An Elman approach. *IEEE Transactions on Intelligent Transportation Systems.* doi: 10.1109/TITS.2021.3088862.

Zhang, J. & Sun, Z. (2016). Assessing multi-hop performance of reactive routing protocols in wireless sensor networks, *Proceedings of IEEE International Conference on Communication Software and Networks (ICCSN).* pp. 444–449.

Zorkany, E. M. (2020, June 9). Vehicle To Vehicle "V2V" Communication: Scope, Importance, Challenges, Research Directions and Future ~ Fulltext. Benthanopen. https://benthamopen.com/FULLTEXT/TOTJ-14-86

Part IV

Use Cases

10 A Case Study on the Smart Streetlighting Solution Based on 6LoWPAN

Manoj Kumar, Prashant Pandey, and Salil Jain
STMicroelectronics
Noida, India

CONTENTS

10.1 Introduction 223
10.2 Architecture of the Smart Streetlighting System 224
    10.2.1 Lighting Module 224
    10.2.2 Sensor Module 228
    10.2.3 Processing Module 228
    10.2.4 Communication Module 229
    10.2.5 System Description 229
    10.2.6 Features of a Smart Streetlight System 230
    10.2.7 Data Concentrator Unit (DCU) Module 231
    10.2.8 Central Management System (CMS) Module 231
10.3 Analysis on Implementation Detail 233
10.4 Conclusion 237
References 237

10.1 INTRODUCTION

Bringing smartness to a traditional infrastructure like streetlight can achieve major cost advantages, by optimizing the energy usage and thus, by extension minimizes the environmental impact of developing such infrastructure. A streetlight installation can be called a "smart" system; if it's efficient, it can take decisions based on its operating environment and it is connected to a network to receive commands and send telemetry data. Such a system typically uses a short-range communication technology to connect each node to a gateway using either a star or mesh network topology. The gateway node in the network helps connect the local nodes to the Internet. A gateway is needed because it's not optimal in terms of cost and resources to connect each node directly to the Internet.

    The smart streetlight system described here is installed inside the STMicroelectronics, Greater Noida campus (near New Delhi, India) and is working flawlessly since many years. This location receives very good sunlight throughout the year barring some days due to cloudy weather and fog conditions. The objective of creating a smart streetlighting system inside the campus is to switch to a renewable power source and to serve as a pilot implementation. This is a hybrid system that charges it's batteries during the day time using the solar panels and depletes the batteries during the night to power the LED light. Additionally, the system can switch over to mains power to charge the batteries when sufficient sunlight is not available. *IPv6 over Low-Power Wireless Personal Area Networks* (6LoWPAN) operating in the sub-GHz band is chosen as communication medium between the streetlights as it allows longer range of operation and can be used for low-power devices with limited processing capability. The 6LoWPAN network has a special node called a border router which acts as a gateway to connect to a central server through cellular network. However, each of the lights is able to work independently even when there is a disruption of the local 6LoWPAN network or the cellular connectivity. The antenna placement is one of the key elements that require attention, considering the outdoor usage and the fact that off-the-shelf casing for a streetlight has primarily metallic enclosure, which limits the radiated RF power availability to the nodes. Radio performance analysis in outdoor scenarios is a complicated issue which heavily depends on the complex of the environmental factors, like the fundamental degradation due to multipath components, reflection, refraction, and scattering (Azpilicueta et al., 2014). An external antenna placed on the pole could also be used instead of the internal antenna, however it may reduce the reliability of the system and increases the installation and repair costs. The free ISM band in India for sub-GHz allows operation in 865 MHz band (TEC, 2015).

10.2 ARCHITECTURE OF THE SMART STREETLIGHTING SYSTEM

To design a modular and expandable system, it is important to give a thorough consideration to the architecture. The smart lighting system consists of three entities communicating with each other—light nodes, data concentrator unit (DCU), and central management system (CMS). Light nodes provide the illumination and sensing functionality in the system while DCU and CMS extend this functionality, making the system truly smart. DCU is a special node in the smart streetlight system that manages the network of light nodes, collects data from the network, and acts as a gateway to sends the data over the Internet to a backend system (Arumugam et al., 2020; Pillai et al., 2020). Backend system is also called CMS or command and control center (CCS) which can communicate to hundreds or even thousands of DCUs.

Each lighting node in a smart streetlight system contains lighting, sensor, processing, and communication modules (Figure 10.1). Each of these modules is discussed in detail below.

10.2.1 Lighting Module

This module provides the core functionality of a streetlight system and has several components (Figure 10.2) like a high-efficiency LED driver, a battery management system, and a solar module. Readers should note that the battery management system and solar module are not essential components in this architecture. A smart

A Case Study on the Smart Streetlighting Solution Based on 6LoWPAN

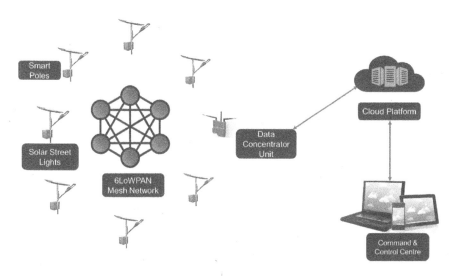

FIGURE 10.1 Architecture of smart streetlighting. (From TEC, 2017.)

streetlight system is expected to use high-efficiency LEDs that require dedicated driver circuitry for optimal operation (Arumugam et al., 2020). LED drivers are basically constant current power suppliers while more sophisticated drivers also consider LED aging and operating temperatures. Some LED driver may support PWM-based dimming (Jha & Kumar, 2019). It's important to do switching at higher frequencies to avoid flicker. Quasi-resonant topology is generally used in the LED driver design as it's efficient and cost effective. If the light is installed in a far-flung place, it's also useful to have a set of solar panel and batteries. An MPPT charger is required to charge the batteries from the solar panel during the daytime, so that lights can run on battery during the night time (Hiwale et al., 2014). Lighting module is generally located inside the enclosure of the LED lamp; however, in some implementations, it may be present inside a separate enclosure. The enclosure including the LED lamp assembly is also called a light fixture, or a luminaire. In more compact designs, LED driver, sensor module, processing module, and communication module, etc., may be mounted on the same PCB. This approach can reduce cost and space requirements; however, careful design is needed to ensure that the electronics are not affected by

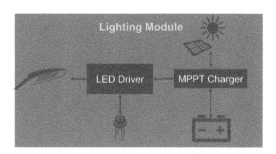

FIGURE 10.2 Lighting module.

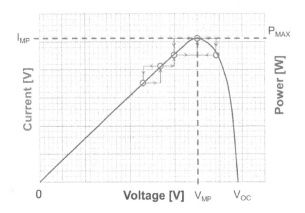

FIGURE 10.3 MPPT working principle. (From www.st.com AN3319.)

the heat generated by LEDs and the Radio Frequency (RF) signal is not obstructed by metallic parts.

Maximum power point tracker (MPPT) charger: It is a DC-to-DC converter connected between solar panels and the battery to regulate the battery charging process. It keeps the voltage and current at an optimal level to extract maximum power from the Photovoltaic (PV) panel, which increases the battery efficiency by 20–45% as compared to non-MPPT controllers.

As shown in Figure 10.3, red line is showing the current-voltage characteristics of a PV panel, whereas the blue curve represents the power transfer. The peak occurs at point V_{MP} and I_{MP}. It can be calculated with the help of Perturb and Observe (P&O) technique (Femia et al., 2004). In this technique, a small perturbation is introduced, which causes the power variation to the PV panel. The PV output power is measured periodically and then compared with the previous one. The same process is continued until the output power increases, otherwise perturbation is reversed. The PV panel voltage is increased or decreased to check whether the power is increased or decreased. It continuously monitors the battery and PV voltage. When an increase in voltage leads to an increase in power, this means the operating point of the PV panel is on the left of the maximum power point (MPP). Hence, further perturbation is required toward the right to reach MPP. On the other hand, if an increase in voltage leads to a decrease in power, this means the operating point of the PV panel is on the right of the MPP and hence further perturbation toward the left is required to reach the MPP. After measuring the battery voltage, it stops charging to prevent over charging, if the battery is fully charged. And if the battery is not fully charged, it starts charging by activating the DC/DC converter. The microcontroller will then calculate the existing power $P_{current}$ at the output by measuring the voltage and current and compare this calculated power to the previous measured power P_{prev}. If $P_{current}$ is greater than P_{prev}, the PWM duty cycle is increased to extract maximum power from the PV panel. If $P_{current}$ is less than P_{prev}, the duty cycle is reduced to ensure the system to move back to the previous maximum power. This algorithm is easy to implement, with low cost and high accuracy.

A Case Study on the Smart Streetlighting Solution Based on 6LoWPAN

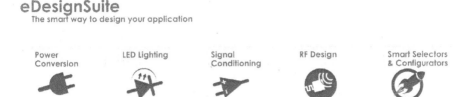

FIGURE 10.4 Designing using eDesignSuite. (From www.st.com.)

To illustrate this point, consider a solar panel that has $V_{MP} = 18.4$ V and $I_{MP} = 11.45$ A. The output of this panel should be ~210 W. A 12-V battery may generally have 12.2 V. In case of a non-MPPT charger controller, the battery would be charged by 11.45 A * 12.2 V = 140 W. Whereas for MPPT, it boosts the voltage and current to MPP of I-V curve of a module as mentioned above. It means battery will charge at 18.4 V * 11.45 A = ~210 W. Therefore, the efficiency increases by >30% in case of MPPT.

It's not easy to design a new power converter from scratch, thus many companies provide tools that help in the design. As an example, **eDesignSuite** (Figure 10.4) is a collection of online software tools provided by STMicroelectronics to simplify and speedup the electronic design process. With the help of eDesignSuite, it is easy to select products and topologies for various types of application like power conversion, LED lighting, signal conditioning, and RF design. It can easily generate the complete set of design documents for the solution which includes fully annotated and interactive schematics, complete bill of materials, current and voltage simulations, efficiency curves, bode stability, etc. (Figure 10.5).

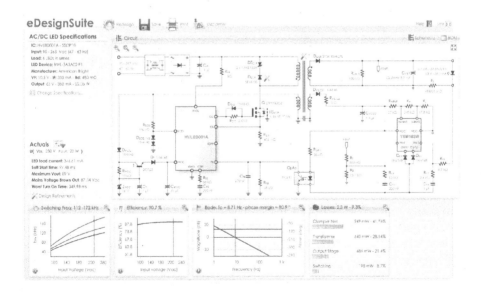

FIGURE 10.5 Smart simulator and system design view. (From www.st.com.)

10.2.2 Sensor Module

Inclusion of a sensor fulfills one of the prerequisites of smart devices owing to its ability to sense the environment. While it is possible to have several sensors installed on a single smart pole, some sensors like ambient light and temperature sensors are essential for the operation. An ambient light sensor allows automatic control of power on/off feature and brightness of the lights based on ambient light levels. Temperature sensors is used to monitor enclosure temperature to prevent overheating of driver electronics and LED heatsink temperature to ensure that the LED temperature does not exceed the design limits, thus helping in extending the LED life. Other useful sensors that can be present on a smart pole are listed below.

- Presence Sensor: It is used to dim or switch off the lights when no one is around. It can also enable additional use-cases like detection of unauthorized entry. Most used technologies for presence detection are: Passive Infrared (PIR), which uses a pyroelectric element to detect infrared emitted by a human body and radar-based motion detectors which emit radio waves and look for changes in frequency of reflected signal (Doppler shift).
- Pollution Sensors: These sensors detect the concentration of various pollutants like CO_2 and SO_2. Many variants of these sensors are based on electrochemical techniques. One class of gas sensors contains metal oxide semiconductor (MOS), which has current flowing through it. When it encounters an oxidizing or reducing gas, its resistance changes, which is interpreted by its electronics. Another class of sensors measures PM 2.5 or PM 10 particles in the air using optical sensing techniques.
- Traffic Sensors: These can be Doppler/radar-based sensors which detect movement of vehicles or acoustic sensors, which detect noise from the passing vehicles. Traffic sensor can be used to control traffic lights further up the road.

Additionally, smart pole may host information displays and public address system. A full suite of sensors is not required on every pole as they can be installed scatteredly across an area skipping several poles.

10.2.3 Processing Module

Processing module can be considered the brain of the system. It's responsible for coordination between various other modules and decision-making. It can also perform limited signal processing and run local analytics to process the data before sending to DCU. Lighting nodes typically use a micro controller unit (MCU) for its processing requirements. MCUs are self-contained computers that include peripherals, communication IPs, RAM, and flash memory on the chip itself. It is possible to extend further the capability of a microcontroller by connecting external RAM, flash chips, or other peripherals. MCUs can be run without an operating system (OS) or can use a real-time operating system (RTOS) to handle more complex applications.

DCU nodes may have higher processing requirements depending on the number of connected nodes and features e.g., local processing capabilities. Thus, a DCU node either employs a powerful MCU as its processing unit or can use even higher processing power and expandability of an industrial microprocessor unit (MPUs). MPUs need external RAM and storage like embedded multimedia controller (eMMC) or flash memory. This can result in higher complexity and design cost; however, MPUs use embedded Linux, thus have access to richer software ecosystem.

There are several popular architectures in use by various MCUs raging from classical 8051 to more advanced AVR, PIC, MSP, and ARM. ARM architecture is by far the most popular architecture in use for 32-bit microcontrollers due to huge software and tools ecosystem and easier vendor portability.[1] ARM processors come in three series – Cortex-A series to be used as general purpose application processors, Cortex-R series for real-time applications e.g., automotive applications while the Cortex-M series covers microcontrollers suitable for embedded applications. ARM Cortex M series microcontrollers are used for the light nodes in a smart streetlight system, whereas ARM Cortex A series microprocessors are recommended for DCU implementation. Another architecture gaining momentum is RISC-V. Although still not in the mainstream, it may witness wider adoption due to its open-source nature of Instruction Set Architecture (ISA).

10.2.4 COMMUNICATION MODULE

A smart light must be able to communicate with other devices to be called a smart device. To keep the cost down and complexity low, it is not possible to host a full TCP/IP networking suite on each node, thus direct connection to the Internet is not possible. The nodes must use a local short-range networking technology to communicate to a gateway node (DCU) that would provide connection to Internet. For local networking, there are various stacks to choose from, e.g., Thread, ZigBee, Bluetooth® Mesh, and 6LoWPAN. To connect the DCU to the Internet, there are multiple long-range communication technologies available to choose from. Technologies like LoRA, Sigfox, and NB-IoT are suited for supporting low data rate applications, whereas traditional cellular technologies like 4G-LTE are best match for high data rate applications.

The data throughput requirement for the DCU-CMS link is dependent on several parameters and sampling rate used for the telemetry data on each node. Data requirements also vary based on the number of nodes generating data inside the network and on how much local data processing is being done by the DCU.

10.2.5 SYSTEM DESCRIPTION

The implementation of schematics and major blocks is shown in (Figure 10.6). The battery can be charged through solar MPPT or direct AC mains when battery voltage falls below a threshold due to inadequate solar feed. Solar charge gets the priority, and a microcontroller takes care of these decisions.

During development, the communication interface uses Sub-GHz to transmit the information about the voltage levels of battery, solar panel, and radio signal

---

1. The Arm Ecosystem: More than just an ecosystem, it's oxygen for SoC design teams, https://www.arm.com/company/news/2019/04/the-arm-ecosystem-more-than-just-an-ecosystem, Accessed 14 Feb 2021.

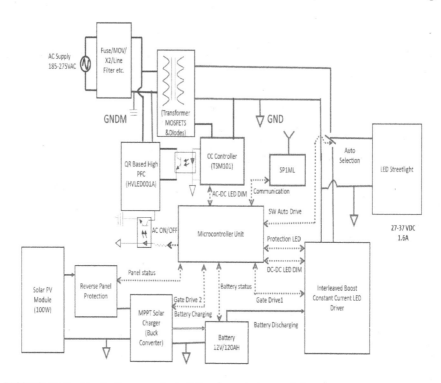

FIGURE 10.6 Block diagram of sub-GHz connected streetlight.

strength received by each node to be displayed on a graphical user interface (GUI). These aspects are important for the development stage. For example, RF conditions are dynamic, and this graphical user interface offers good insights into development and debug problems, including node discovery, rejoining, authentication, etc. In production, various parameters reported by the system help to determine if the system is running normally and there are no connection or signal issues.

10.2.6 Features of a Smart Streetlight System

The smart streetlight system has following features:

- Electronic power and intensity control to reduce power consumption
- Telemetry, status, and health check reporting
- PIR and other sensors for added smartness
- Quasi-resonant driver for mains operation
- Efficient solar battery charger (with MPPT)
- Protections like panel reverse and LED open/short

It is important to mention here that these details are meant to give a view of the details that are needed from practitioners' point of view. In an actual installation, all the features may not be present. It is also possible that in an installation, majority of the lights have basic features, whereas a few of the lights in the same installation may have extra features. Such variations in the requirements within an installation and across different installation highlight the need to have a modular system design so that set of modules may be customized for each pole as per the requirements.

Another important feature of a smart streetlight system is the condition monitoring and fault reporting capability. Condition monitoring is a feature which can warn about an impending failure and thus facilitating preventive maintenance of the system. For instance, if the system observes that an LED module has started to draw more than usual current while operating at some illumination level, or the temperature of an enclosure is too high compared to other lights with same configuration, it is possibly an indication of impending failure. If a failure does occur, the smart light should inform the maintenance team through the CMS, so that the failure can be fixed even before users of the system make a complaint.

10.2.7 DATA CONCENTRATOR UNIT (DCU) MODULE

Primarily, the DCU provides Internet connectivity to the lighting nodes. The DCU uses two communication modules—one to communicate with the lighting nodes using a short-range communication protocol like 6LoWPAN and the other one to communicate to the backend using a long-range communication technology. In this context, DCU is responsible for managing the local network of light nodes. It is also responsible to bring in newly installed nodes into the network, while making sure that unauthorized nodes are rejected. DCU also manages the network parameters and configuration of local nodes. These settings can be updated by DCU based on the commands received from the CMS. DCU is also instrumental in updating the firmware of the light nodes when a new firmware image is received from the CMS. In case of loss of connectivity with the CMS, the DCU can cache important messages and send them to the CMS when the connectivity is restored.

DCU can perform local processing of data to save the bandwidth on the CMS link, to reduce latency, and to reduce load on CMS. The local processing can be of two types: one is to take decision locally, e.g., operate light nodes based on input from sensors, other one is to digest the data by performing filtering and data aggregation on the telemetry data before sending it to the CMS. This functionality is popularly known as edge processing.

Another important role of the DCU is to act as a firewall to protect nodes in the local network from outside threats. DCU communicates with the CMS on an authenticated and encrypted link. The DCU also sanitizes the data being received from the internal light nodes before sending to the CMS. In other words, DCU acts as the gatekeeper for the network of light nodes.

10.2.8 CENTRAL MANAGEMENT SYSTEM (CMS) MODULE

The primary function of the CMS module is to collect telemetry data from the nodes. Additionally, it can send command to the nodes and perform remote maintenance, for example, updating configuration of the nodes or sending firmware updates. Note that all the communication of the nodes with the CMS happens through the DCU.

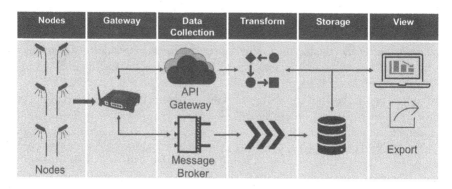

FIGURE 10.7 System view of streetlighting.

CMS is basically a *software component running on server*. The software can run on a self-hosted server or a cloud-based server. If the streetlight system is large, it is appropriate to host the CMS on a cloud service provider like Amazon AWS or Microsoft Azure. The CMS collects telemetry data from multiple DCUs and presents it to the users on a dashboard for quick understanding of the overall situation. It is possible to apply analytics on the telemetry to uncover underlying patterns and alert users about any anomaly present in the system. Various components of the CMS are discussed below (Figure 10.7).

- *API Gateway*: An API gateway is a secure entry point to the CMS; it exposes endpoint for the DCU to make an API call. API gateway also authenticates the DCU to make sure that a malicious device does not connect to the network. API gateway works on a protocol like HTTPS. For supporting real-time protocols like MQTT, a message broker is employed. Message broker is similar to the API gateway except that it works on a pub-sub configuration (Singh et al., 2015), thus facilitating intercommunication between several DCUs.
- *Ingestion Pipeline*: Large streetlight installations may generate huge amounts of data. Ingestion pipeline includes a queue to buffer this data before it can be processed (Meehan & Zdonik, 2017). Furthermore, the pipeline may filter the data stream based on various parameters and deliver the data points to different components of the processing pipeline.
- *Processing Pipeline*: Processing pipeline includes components to transform the data using stream processing techniques. The transformed data can be used by upstream modules in the CMS and at the same time, this data can also be stored in a No-SQL database for batch processing. Processing pipeline may have configurable triggers which can initiate an action based on a pre-set criterion including sending notification or mails to the administrator.
- *Visualization*: Visualization is a very important component of a CMS, where the real-time data is shown on a dashboard. It helps the human operators administering the system to get a bird's-eye view of the entire system and react if a problem occurs.

CMS can also be provided as a subscription-based service where data packets are sent to the cloud and analyzed for anomalous behavior.

10.3 ANALYSIS ON IMPLEMENTATION DETAIL

Here, we describe the software implementation details of the solar smart streetlight solution installed in the premise of STMicroelectronics. There are 40 light nodes installed in a designated parking area controlled by a DCU that further connects to a simple CMS specifically developed for the purpose. The wireless communication between the light nodes and DCU uses 6LoWPAN mesh technology. The system works in the 865–867 MHz frequency band which is within the designated ISM frequencies in India (TEC, 2017). The solution described here uses 6LoWPAN over sub-GHz, given that it is open source and stable. Furthermore, it is supported natively in Linux kernel, and OS like Contiki uses it as a primary network stack (Yang & Chang, 2019). The network has a mesh topology, so that any node can talk to DCU through another node.

The communication between DCU and CMS uses 4G-LTE module. The CMS allows the user to switch on/off or dim individual lights or group of lights. The user can also issue time-based commands through the CMS. The CMS interface displays different electrical parameters for each light node along with its status. If a node develops a fault or goes out of network, that information is shown on the web interface. All the telemetry data is stored on the server and can be used for data analysis. Figure 10.8 shows the interface of the sample CMS application.

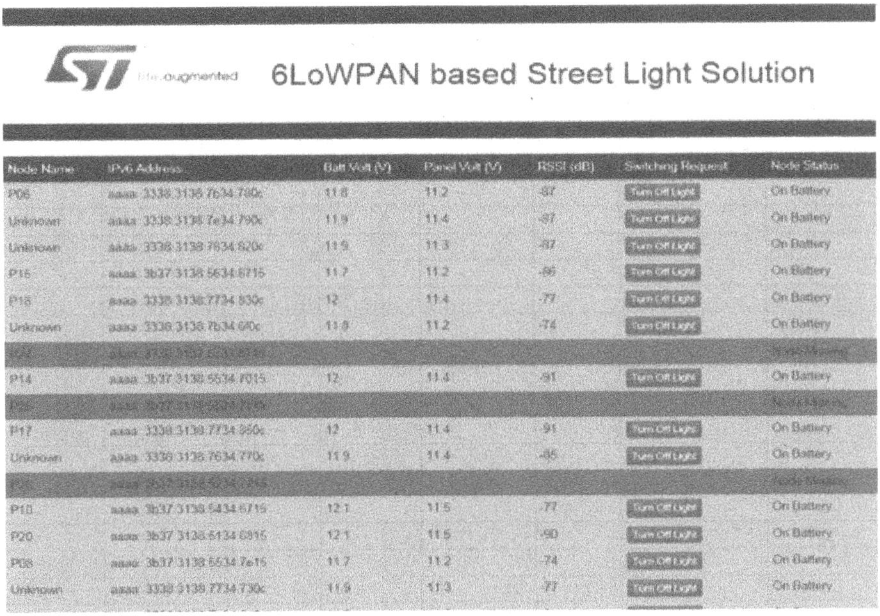

FIGURE 10.8 Cloud analytics. (From TEC, 2017.)

The CMS described in this section has been implemented in ASP.NET and hosted on Microsoft Azure platform. The entire CMS system is implemented using following modules.

1. Data Collection Module: This is also called an ingestion module and it provides an API endpoint for the clients (DCUs) to send telemetry data to the CMS. DCUs can connect to the module either using HTTPS protocol or using MQTT protocol. HTTPS allows secure connections from the DCU to the server. However, it follows a request-response model. Thus, if the CMS has a command to send to the DCU, it must wait until a request is initiated from the DCU side. CMS can then send the command in response to that request. HTTP may terminate underlying TCP connection once the request is completed. MQTT, on the other hand, maintains the underlying TCP connection, thus any side can send the data at any time. It can be deduced that the HTTPS protocol results in higher latency but needs less server resources (fewer connections to maintain). MQTT has much lower latency but requires the server to maintain a TCP connection for each connected DCU, thus requiring extra resources from the server (data collection module). MQTT has additional features like ability to define "topics," which are analogous to channels. Any node can subscribe to one or more topics or publish to a topic. This allows for a robust multicast messaging mechanism. This module also acts as the first line of defense against cyber-attacks. It authenticates and confirms identity of the connected DCUs based on either device certificates or API keys. Only authorized DCUs can connect and transfer data. Message queuing is another feature of this module. It can store the incoming messages in a queue until they are processed by the next module. Data ingestion functionality is offered as a managed service by most of the popular cloud providers, e.g., Azure Event Hub. There exist managed cloud services such as Azure IoT hub service and AWS IoT core, which provide full IoT communication functionality right from ingestion to authentication and messages queuing to processing. Software like Rabbit MQ and Apache Kafka can be used by implementers to create own data ingestion pipelines. Factors like load balancing and distributed denial-of-service (DDoS) attack protection should also be considered during the design phase.
2. Processing Module: The processing module performs real-time as well as batch processing on the telemetry data. Some amount of real-time processing is always required, for example, the incoming data must go through basic filtering and transformation before saving it to a database for later processing. Also, some processing must be done in real time to detect parameter's threshold breaches or other alarming conditions, which must be handled or notified immediately. On the other hand, many computation-intensive operations like updating machine learning models or generating reports are mostly run in batch-processing mode. The real-time processing of telemetry data is done using stream-processing tools. Kafka is a popular open-source software that provides decent stream-processing capabilities while Hadoop and Apache spark are heavyweight tools with extensive stream processing, batch

processing, and storage capabilities. On the managed services front, Azure stream analytics and Amazon Kinesis are popular cloud-based stream-processing solutions. To store the raw data, a blob-based storage like Amazon S3 or Azure Blob can be used, and to store the processed data, a solution like Azure Cosmos DB or AWS Dynamo DB can be used.

3. Visualization Module: The module represents a web application that provides an interface between CMS and a human operator. The purpose of this module is to present data in a visually simple to understand manner as well as allow operator to configure the system and send commands. It consists of a backend (a web server) and a frontend (a web client). The application described in this case study uses a managed ASP.NET application running on an Azure App Service as the backend. Node.JS is an alternative framework that can be used to provide functionality similar to ASP.NET. The frontend for the case study is developed using standard web stack—JavaScript, HTML5, and CSS3. Bootstrap is used as a layout engine to ensure that the UI scales well to different screen sizes. Tailwind CSS is a viable alternative if the readers want to use a different layout system. The contemporary way to create a data-rich web client is to use a frontend JavaScript framework like React or Vue for state management of the application. Though it is not the part of the case study, readers are encouraged to use such a system. The web client (frontend) communicates with the backend using restful HTTPS APIs with all data getting exchanged in the JSON format, which is a standard format used by APIs to transfer data. JSON has largely replaced XML as it's more concise and expressive. If real-time response is required, HTTP can be replaced with a WebSocket as a communication protocol. WebSocket and MQTT both use TCP as underlying transport, but WebSocket handles firewalls and proxies much better compared to MQTT due to the use of ubiquitous port 80. MQTT over WebSockets is a mechanism which allows MQTT to use WebSocket as an underlying transport.

There are three models through which CMS functionality can be offered. The first is Software as a Service (SaaS) model. In this model, full CMS software is provided on a subscription basis, e.g., StreetMan from Dhyan, LightingGale from SIMCON, and SLV-CMS from Itron. This model is the fastest to deploy as no development is required; however, avenues of customization are limited and a recurring fees must be paid to the vendor. The second option is CMS in Platform as a Service (PaaS) model. In this model, the CMS functionality has to be implemented but managed services such as App Service, IoT Service, and Database Service provided by cloud providers are used as a building block. It is relatively fast to setup because configuration, updates, security, and performance tuning are managed by the cloud provider. The third option is to build CMS using Infrastructure as a Service (IaaS) model, using open-source software as building blocks on a virtual machine platform provided by the cloud service provider. This is the most flexible strategy and it's possible to run all the CMS modules on a single virtual machine to cater to a typical office campus while spending under $50 per month. However, all configuration, software upgrades, licenses, etc., need to be managed by the implementer. For larger installations,

FIGURE 10.9 Field installation at STMicroelectronics, Greater Noida, India. (From TEC, 2017.)

however, the configuration and maintenance of the system become challenging and considerations like scaling, security and uptime also become important. In such scenarios, use of managed cloud services is recommended. As these services are billed only for the actual usage, a carefully architected system can deliver considerable cost savings.

This being a pilot project (Figure 10.9), adequate manual overrides have been placed at multiple layers. For example, physical switches are provided at the DCU, in case there is a cellular network failure due to any reason. During the operation of the network, the local network switches to autonomous mode, wherever there is absence of a GSM signal, or the CMS is down for maintenance. The main purpose of this system was to experiment and analyze the behavior of nodes joining, rejoining, authentication, range testing for RF module, antenna orientations, and installation issues. For this reason, it was important to capture large number of parameters of each of the node during the experimentation phase. Once a system is installed, these details are no longer useful, thus a simpler interface can be designed. This strategy could be used by readers as well when designing their own smart streetlight system.

The system is designed to prioritize the use of solar energy. The battery gets charged with the help of a solar panel in the daytime. During the night time, stored energy is used to power the lights. The system uses the mains power as a backup source, in case the battery is not fully charged due to bad weather or some other reasons. As the ambient light goes down, the voltage generated by the panel also decreases which triggers to switch on the lights. The lights automatically switch off as the sun rises and solar panel starts generating output, thus the solar panel also serves as an ambient light sensor. The lights also support a power saving feature dimming the LEDs during late night hours, when no traffic is expected. This feature is enhanced by adding a presence sensor to restore full brightness as soon as some movement is detected.

10.4 CONCLUSION

The savings as a result of these installations can be calculated by comparing the metering data of the smart lighting installation with the historic data of previous installations. The upfront costs of solar hybrid systems are higher but can be offset within a period of 2 to 3 years due to better operational efficiencies, however, the payback period also depends on the energy prices of the location. The benefits of cloud connectivity go beyond cost saving. As mentioned earlier, a smart pole integrates a plethora of sensors and multiple communication technologies working together with a cloud-based application to make intelligent decisions. This enables very high operational efficiency and advanced use cases, especially for smart cities. However, smart lighting systems also come with new challenges. These systems require more expertise for manufacturing, installation, and configuration. Connected lighting system may be vulnerable to cyber-attacks by malicious parties sitting across the globe. Thus, security becomes one of the key parameters that must be kept in mind while designing such a system.

REFERENCES

Arumugam, P., Jha, A. K., & Kumar, M. (2020). *Evaluation of Heterogeneous Last Mile Technology for Smart Metering and G3-PLC Based Smart LED Streetlighting Application. January.* https://doi.org/10.1007/978-981-32-9119-5

Azpilicueta, L., Falcone, F., Astráin, J. J., Villadangos, J., Chertudi, A., Angulo, I., Perallos, A., Elejoste, P., & García Zuazola, I. J. (2014). Analysis of topological impact on wireless channel performance of intelligent street lighting system. *Radioengineering, 23*(1), 412–420.

Femia, N., Petrone, G., Spagnuolo, G., & Vitelli, M. (2004). Optimizing sampling rate of P&O MPPT technique. *PESC Record – IEEE Annual Power Electronics Specialists Conference, 3*(July), 1945–1949. https://doi.org/10.1109/PESC.2004.1355415

Hiwale, A., Patil, M., & Vinchurkar, H. (2014). An efficient MPPT solar charge controller. *International Journal of Advanced Research in Electrical, Electronics and Instrumentation Engineering, 3*(7), 10505–10511. https://doi.org/10.15662/ijareeie.2014.0307017

Jha, A., & Kumar, M. (2019). Smart LED driver with improved power quality and high efficiency for household application. *2018 International Conference on Computing, Power and Communication Technologies, GUCON 2018, March,* 248–253. https://doi.org/10.1109/GUCON.2018.8674943

Meehan, J., & Zdonik, S. (2017). Data ingestion for the connected world. *Cidr.* https://cs.brown.edu/courses/cs227/papers/brown-data-ingest.pdf

Pillai, R. K., Seethapathy, R., Sonavane, V. L., & Khaparde, S. A. (2020). *ISGW 2018 Compendium of Technical Papers* (Vol. 580, Issue November). https://doi.org/10.1007/978-981-32-9119-5

Singh, M., Rajan, M. A., Shivraj, V. L., & Balamuralidhar, P. (2015). Secure MQTT for Internet of Things (IoT). *Proceedings – 2015 5th International Conference on Communication Systems and Network Technologies, CSNT 2015,* 746–751. https://doi.org/10.1109/CSNT.2015.16

TEC. (2015). *Technical Report M2M Enablement in Power Sector, Spectrum Requirements for PLC and Low Power RF Communications.* http://tec.gov.in/pdf/M2M/M2M Enablement in Power Sector.pdf

TEC. (2017). *Technical Report: Communication Technologies in M2M/IoT Domain, TEC-TR-IoT-M2M-008-01.* https://tec.gov.in/pdf/M2M/Communication Technologies in IoT domain.pdf

Yang, Z., & Chang, C. H. (2019). 6LoWPAN overview and implementations. *Proceedings of the 2019 International Conference on Embedded Wireless Systems and Networks,* 357–361. https://dl.acm.org/doi/10.5555/3324320.3324409

11 IoT-Enabled Real-Time Monitoring of Assembly Line Production

Maneesh Tewari and Devaki Nandan
College of Technology
Pantnagar, India

CONTENTS

11.1 Introduction 239
11.2 Literature Review 240
11.3 Research Methodology 242
    11.3.1 Existing Situation...242
    11.3.2 Proposed Methodology..242
    11.3.3 Components of the System 242
        11.3.3.1 Front-End Edge Devices 243
        11.3.3.2 Microcontrollers..246
        11.3.3.3 Code Developed 246
        11.3.3.4 Circuit Used 247
    11.3.4 Testing Procedure in Industry 249
11.4 Results After Testing 251
11.5 Discussion..252
11.6 Limitations and Challenges of Project 253
11.7 Future Scope..253
11.8 Conclusion 254
Acknowledgment 254
References..255

11.1 INTRODUCTION

Cycle time is an important aspect to attain business efficiency. Time study is applied to measure the standard processing time. For a multi-station assembly line, the highest time required on any of the workstations is taken as the cycle time of the whole line. Thus, cycle time decides the speed of production. The problem arises when the line doesn't work on the standard cycle time, which results in less production than desired.

Most of the companies are not able to achieve the daily production target even if the line balancing is already done, which reflects the inefficiency of the production line. This is the result of not following the standard cycle time established by the time study. However, this can be avoided by effective monitoring of the cycle time

DOI: 10.1201/9781003122357-15

at each workstation on the assembly line. If the management is aware of the number and amount of cycle time breaches, delays, and stoppages on each workstation at regular intervals, then they can look into the reasons and take preventive measures timely and avoid the problem of underproduction. Getting this information manually is a tedious and time-consuming task. By the time results are compiled manually, a significant amount of shift is over and the manager cannot do much about the underproduction. This problem can easily be overcome by real-time monitoring of the assembly line.

If the management is able to detect the variation in cycle time on a real-time basis, then they can take effective measures to ensure that the line runs at standard cycle time. In this project, a monitoring system was developed for cycle time on an assembly line so that the management is aware of when and where cycle time is breached and how many times in a particular shift. The management also gets information about the parameters such as idle time, balance delay, and projected target on a real-time basis. Apart from talking about cycle time breaches, the system also recognizes inefficient workers.

The system proposed by us in this project employs Internet of Things (IoT), which eliminates manual work for monitoring the assembly line and leads to a more efficient assembly line to achieve the daily production target. Thus, the objectives of the system developed are as follows:

1. Effective monitoring of the assembly line on a real-time basis.
2. Computing actual cycle time of each workstation on a real-time basis.
3. Identification of bottlenecks on a real-time basis and enabling managers to make decisions to work upon them.
4. Keeping a record of data related to cycle time breaches and delay at each workstation.

11.2 LITERATURE REVIEW

An assembly line is basically a combination of man, machine, and material in a manufacturing process where parts either raw or semi-finished state are fed at a constant rate from one point to another, called workstations, in a logical way to get a final product in the most predictable way (Shukla et al. 2018). Assembly Line Balancing (ALB) is defined as the optimum allocation of tasks to different workstations to have maximum utilization of manpower at a workplace. The purpose of this type of task allocation of each worker is meant to achieve a balanced assembly line with maximum productivity (Kumar and Mahto 2013). The total processing time at each workstation, as a rule, never exceeds cycle time for smooth production output (Roshani and Giglio 2017, Taifa and Vhora 2019).

A robust monitoring system is an essential element in a production setup. It not only keeps track of the progress but plays a key role in cost cutting (Zhang et al. 2017) and providing an early warning system (Manes et al. 2016, Cheung et al. 2018). Modern monitoring systems comprise IoT-based sensors (Syafrudin et al. 2018). Various studies conducted for reducing industrial fatigue have emphasized on using these types of sensors for production monitoring (Moon et al. 2017), error design prevention (Salamone et al. 2017), fault diagnosis (Li et al. 2018), quality prediction

(Lee et al. 2018), and helping managers with better decision-making (Calderón Godoy and González Pérez 2018).

IoT technology can be implemented for ensuring real-time data capturing and monitoring, thus helping in a smooth production line (Xu and Chen 2016). IoT applications have now been accepted in sectors such as healthcare, logistics, supply chain, manufacturing, pharmaceuticals, etc. The extended application of IoT – called IIoT is now part and parcel of wireless communication as well as network technologies. The advantages of these technologies include increased safety, automation, performance reliability, etc., in industrial sectors (Hou et al. 2019, Abuhasel and Khan 2020, Khan and Abuhasel 2020, Khan and Alghamdi 2021). IIoT is basically connecting people with embedded technologies (Kortuem et al., 2009). In IoT, information gathering and exchange by the common man almost become the next generation of services. Each nodal point in the chain starts playing the role of sensors and actuators (Gubbi et al. 2013). The communication between various devices in the system can be made by many available technologies (Ebling and Want 2017). In an ever-changing dynamic manufacturing environment owing to changing customer needs, IoT technology simplifies the production scheduling as inventory, such as raw material, semi-finished as well as finished goods together with man hours, machine hours, etc., is linked together (Xu and Chen 2018).

Subramaniam et al. (2007) worked on the need for knowing the status of information on a real-time basis at every stage of the production line so that quick feedback system could be developed for precision manufacturing, which leads to higher productivity. Borra and Shinde (2017) proposed a new technique using IoT to effectively transfer the data by use of electronic update trackers. The implementation of the system included the introduction of slave trackers throughout the assembly line using IoT. After each assembly in the system, the worker feeds the information to the system, and the slave tracker is updated to transfer further updates to master tracker and it continues till the final assembly is over. Any deviation from the target or gap in demand is easily monitored by the supervisor and the necessary corrections are immediately made. This technique eliminates the need for manual data update and frequent line stoppages.

Syafrudin et al. (2017) proposed a real-time sensor-based IOT for the factory assembly line. The Raspberry Pi 3 model b (2) was used as an IoT device with several sensors, that is, temperature sensor, humidity sensor, ambient light sensor, air quality sensor, gyroscope, accelerometer, etc. The IoT device will send the collected data into the system every 5 seconds. Once the worker is performing the task for the assembly of door trim, the IoT device will start feeding the sensor data into the system. The system then will process and classify the data and show the result on a real-time monitoring page. Thus, this system displays data related to temperature, humidity, light, and air quality in the assembly chamber.

Ruppert and Abonyi (2018) developed an IIoT-based algorithmic solution so that products and assembly lines are connected on a real-time basis for an error-free production setup. The wireless sensors played a key role along with Indoor Positioning System (IPS) for continuous resetting of cycle time due to changing demands on the line. This not only eliminated human intervention but introduced automation of feedback of information in a continuous production line.

11.3 RESEARCH METHODOLOGY

11.3.1 Existing Situation

In most of the companies, either monitoring of cycle time is neglected or done manually. As cycle time is the measure of production line's speed and efficiency, neglecting the problems related to cycle time or knowing them late results in less productivity. The movement of bottleneck is very difficult to identify manually. Cycle time is the time taken at bottleneck workstation and is established once; however, in actual cases, the bottleneck shifts from one station to another, thus cycle time of the line also changes. This change in the cycle time results in varying production speeds. Identifying and removing the shifting bottlenecks in mass production is next to impossible if done manually. So, a real-time monitoring system, if employed, helps in eliminating the shortcomings of a manual system.

11.3.2 Proposed Methodology

A flow chart of the approach used is shown in Figure 11.1. It is used to find actual cycle time on every workstation on the assembly line and if the cycle time is not known on a real-time basis, then effective measures cannot be taken timely. So, sensors are installed at the output stage of each workstation. These sensors will count the number of parts passing through the workstations and send the data to a microcontroller. After a particular pre-set period (say 60 seconds), the microcontroller will calculate the actual cycle time based on this counting and update it on the display. In short, the actual cycle time can be calculated by simply finding the time taken for the counted number of pieces passed through a workstation in a particular time as per the following formula:

Actual cycle time = Pre-set time period/Number of pieces counted

Using a suitable programming language for the microcontroller, the logics can be developed and parameters like delay, projected target, idle time, etc., can be calculated as the actual cycle time is already calculated. These parameters can be displayed for each workstation on excel spreadsheets or display platforms created using third-party communication software.

The results displayed can be monitored by the line supervisor or manager for taking suitable actions whenever required. Figures 11.2 and 11.3 show the schematic diagram of the system proposed.

11.3.3 Components of the System

The approach adopted was to develop a real-time monitoring system by using Arduino which is simple to use as compared to Programmable Logic Controllers (PLCs) using the concept explained in the methodology. For the purpose of demonstration, a model was developed consisting of a single workstation. In this model, an IR proximity/obstacle detecting sensor, Arduino Uno R3 ATmega328P, jumper wires, USB cable, Arduino Integrated Development Environment (IDE) software,

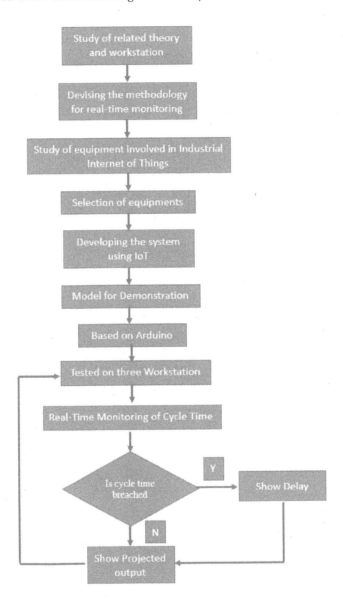

FIGURE 11.1 Flowchart of research methodology used.

and laptop were used for the execution of the program and displaying the data. Apart from demonstration, the model was tested in the said industry on three workstations. A detailed description of the different components used is given below.

11.3.3.1 Front-End Edge Devices

Front-end devices such as control devices and sensors are essential elements of IIoT as these devices are responsible for the continuous collection and streaming of data.

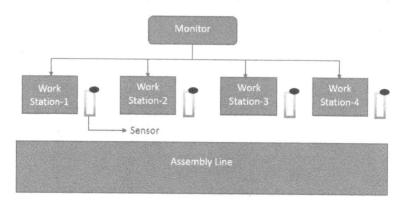

FIGURE 11.2 Schematic diagram for proposed methodology.

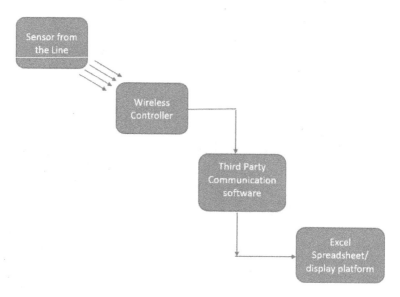

FIGURE 11.3 Schematic diagram depicting the working methodology.

A sensor is a device used for timely detection of any unwanted events or changes in its vicinity and immediately sending the information to the feedback system. A wide variety of sensors and measuring devices are used in the industry, for example, temperature sensor, proximity sensors, infrared sensor, ultrasonic sensor, etc.

In this project, an infrared proximity/obstacle detecting sensor is used (Figure 11.4). This sensor detects any unwanted object from a distance. A proximity sensor has a popular application in various industries that functions on the principle of electromagnetic radiation to detect any obstacle for machine part movement. A capacitive proximity sensor is used for plastic part identification, while an inductive

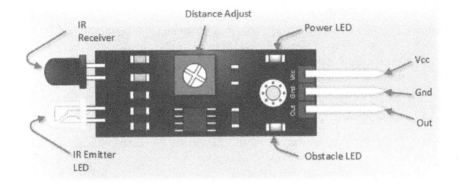

FIGURE 11.4 Labeled picture of an IR Proximity/Obstacle detecting sensor.

proximity sensor is used for metal object identification. This absence of physical contact between proximity sensors and the objects allows these sensors to have high reliability and long functional life.

The package has three connection pins:

1. Vcc to the power supply 3–5 V DC
2. GND to the ground reference
3. OUT for the digital output signal of the sensor

Placing the object in front of above sensor means a logic one (+5 V) at the digital output while no object before a sensor means a logic zero (0 V). Two LEDs are mounted onboard – one for indication of power while the other for indication of object's presence. This digital output can be directly connected to an Arduino or any other microcontroller to read the sensor output. The specifications are written in Table 11.1.

TABLE 11.1
Specifications of IR Proximity/Obstacle Detecting Sensor

Model	Infrared Proximity/Obstacle Detecting Sensor
Input Voltage	4.5–6 V
Operating Voltage	5 V
Number of Pins	3(out, Vcc, GND)
Output Types	Digital Output
Lead Pitch	0.1th inch (2.54 mm) Breadboard compatible
Dimension	15.25 mm × 33 mm × 12.7 mm
Operating Temperature	0°C–60°C (± 10%)
Range	0.5–5 cm
Weight	Approx. 10 g

11.3.3.2 Microcontrollers

A microcontroller is defined as a miniaturized computer on a single integrated circuit. They are specifically made for embedded applications as opposed to a microprocessor, which is generally employed in PC's and other non-tailor-made applications that constitutes of various discrete chips. Microcontrollers are standalone single-chip ICs that contain CPU, ROM for storing program, RAM for storing constants and variables used upon the execution of a program, and different kinds of I/O buses that serve the purpose of connecting to the outside world like UART, SPI, I2C, etc. It cannot execute any program by itself. It must be programmed via an external interface to a PC. They may also require an external crystal for providing a clock, but there are some which already have an internal clock.

For our purpose, we need to use either a microcontroller on some sort of development board or PLC, which is a high-level microcontroller. Both have their own pros and cons. An Arduino being a board/breakout provides a base to the microcontroller chip to facilitate an easy access to analog pins, smooth input/output, and effortless burning/uploading of a program. In this project, Arduino Uno R3 Atmega328P is used.

11.3.3.3 Code Developed

The Arduino language is nothing but a set of C/C++ functions which can be called upon from the preexisting code. There are some minute changes (e.g., automatic generation of function prototypes) which are then directly passed on to a C/C++ compiler (e.g., avr-g++). Arduino supports all standard C/C++ constructs supported by avr-g++. In this project, Arduino IDE is used for programming the Arduino board. However, the Arduino board can also be programed without Arduino IDE.

Arduino Software (IDE) has various features like a text editor meant for writing a piece of code, a text console, a message area, a series of menus, and also a toolbar with buttons for common functions. It connects the Arduino and Genuino hardware to upload programs and also communicate with them.

Following is the pseudo code used in developing the system:

```
ir_pin := 8 //output from sensor to pin 8 of Arduino
counter := 0 //variable to count number of pieces passed
tim[100]
timer := 0
ct := 0 //variable for cycle time
shift_time := 7.5*3600 //considering shift of 7.5 hours
proj_target //variable for projected target
hitObject := false
i := 0

Function setup()
        Serial.begin(9600)
        pinMode(ir_pin,INPUT)
        Serial.print "WORK-STATION: 1"
```

```
Endfunction
Function loop()
        val := digitalRead(ir_pin)
        IF val = 0 and hitObject = false THEN
              counter++
              hitObject := true
              tim[i] := millis()
              timer := tim[i] - tim[i-1]
              i++
              ct := timer/counter
              Serial.print "Cycle time = "
              Serial.print ct/1000
              proj_target := shift_time/(ct/1000)
              Serial.print "Projected Target = "
              Serial.print proj_target

              IF ct/1000 > 4 THEN
                      Serial.print "Delay"
              ELSEIF  ct/1000 = 4 THEN
                      Serial.print "Close enough"

              ENDIF
              counter := 0
        ELSEIF val = 1 and hitObject = true THEN
              hitObject := false

        ENDIF
Endfunction
```

11.3.3.4 Circuit Used

The circuit used in our model is quite simple and is made using components mentioned. The connections are made as follows:

1. The output pin of the sensor is connected to pin 8 of the Arduino
2. The ground pin (GND) of the sensor is connected to the GND of the Arduino
3. The Vcc pin of the sensor is connected to a 5 V pin of the Arduino
4. The Arduino is connected to a laptop via a USB cable

Figure 11.5 shows the circuit used in the model.

A circuit diagram of Arduino is given in Figure 11.6. Following are the pins and their functions (Zait 2018).

Barrel Jack – A barrel jack, or DC power jack, can be used to power an Arduino board. The barrel jack is usually connected to a wall adapter. The board can be powered by 5–20 volts but the manufacturer recommends keeping it between 7 and 12 volts. The regulators might overheat if kept above 12 volts, whereas if kept below 7 volts, they might not perform as desired.

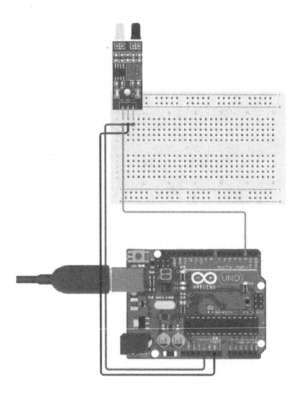

FIGURE 11.5 Circuit used in the model.

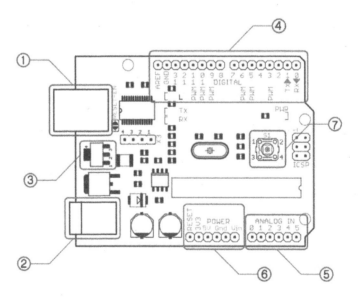

FIGURE 11.6 Pin diagram of Arduino: 1) USB connector; 2) power connector; 3) automatic power switch; 4) digital pins; 5) analog pins; 6) power pins; 7) reset switch.

Vin Pin – This pin is used to power the Arduino Uno board using an external power source. The voltage should be within the range mentioned above.

USB cable – When connected to a computer, it provides 5 volts at 500 mA.

Power Supply – They provide regulated 5 V and 3.3 V to power external components according to manufacturer's specifications.

GND – In the Arduino Uno pinout, you can find 5 GND pins, which are all interconnected. The GND pins are used to close the electrical circuit and provide a common logic reference level throughout the circuit. It should be made sure that all GNDs (of the Arduino, peripherals, and components) are connected to one another and have a common ground.

RESET – Resets the Arduino

IOREF – This pin is the input/output reference. It provides the voltage reference with which the microcontroller operates.

Analog IN – The Arduino Uno has six analog pins, which utilize Analog to Digital converter (ADC). These pins serve as analog inputs but can also function as digital inputs or digital outputs.

Digital Pins – Pins 0–13 of the Arduino Uno serve as digital input/output pins. Pin 13 of the Arduino Uno is connected to the built-in LED. In the Arduino Uno, pins 3, 5, 6, 9, 10, 11 have pulse width modulation (PWM) capability.

11.3.4 Testing Procedure in Industry

Under this project, we tried to test the feasibility of our system at Minda Industries Limited (Acoustic Division), Pantnagar, which is a subsidiary of Uno Minda Group. The problem in the Acoustic Division of the industry, which is indulged in the assembly of the horns of two and four wheelers, is that even their takt time (4.1 seconds) is greater than their cycle time (4 seconds) still they are not able to meet their daily production target.

Uno Minda is one of the leading suppliers of proprietary automotive solutions to original equipment manufacturers. Founded in 1958 with small start-up capital and now notching up a group turnover beyond US$875 million, speaks volumes of the conglomerate that it is today. Uno Minda has 52 manufacturing plants in India, Indonesia, Vietnam, Spain, Morocco, Mexico, and Colombia; design centers in Taiwan, Japan, and Spain; sales offices in USA, Europe, Vietnam, & Spain.

There are several assembly lines in the Acoustic Division of the industry differing based on the diameter of the horns being assembled. The assembly line on which we worked assembles the horns of 70-mm diameter and it consists of 20 workstations.

The cycle time of the assembly line is established by the time study. Once a cycle time is established, it remains the same for a large period of time say one year. Thus, the speed of production is fixed and if the line runs on the established cycle time, the target should be achieved. Even after the employment of this technique, the industry is not able to achieve the daily production target and they are forced to do the overtime, which brings a lot of direct and indirect costs associated with it. So, their existing approach is incapable to identify the root cause of the inefficiency in their assembly line. This is happening because in mass production involving manual work, the cycle time doesn't remain constant. The

FIGURE 11.7 System (shown in red circle) installed after nucleus caulking workstation.

company does not have any system to detect this. So, our system can be of great help in this area.

Since it was a costly and complex affair to implement the system on 20 workstations for the purpose of testing, we implemented our system on 3 workstations, namely, nucleus caulking, resonator riveting, and air gap checking as shown in Figures 11.7–11.10. The selection of these workstations was made based on the fact that these were the workstations with task times very near to the cycle time.

FIGURE 11.8 Horns passed to the next workstation after nucleus caulking.

IoT-Enabled Real-Time Monitoring of Assembly Line Production 251

FIGURE 11.9 System placed after resonator riveting workstation.

FIGURE 11.10 Testing of the system while the line is running.

11.4 RESULTS AFTER TESTING

The system was tested in industry on three workstations, viz., nucleus caulking, resonator riveting, and air gap checking. The results obtained after running the system are shown in Figure 11.11. The system showed three parameters:

1. Actual cycle time on the workstation
2. Projected target with the actual cycle time
3. Delay warning if cycle time is breached

As the cycle time established by the time study does not remain constant, it is subjected to variations even on the same workstation especially when the work is done manually. Three workstations were chosen where task time is close or equal to the cycle time. The system calculated and displayed the actual cycle time on the workstation.

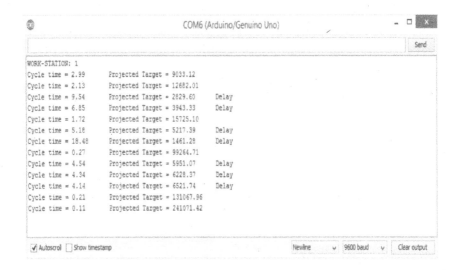

FIGURE 11.11 Screenshot of the serial monitor showing the results.

Minda Industries Limited employs two shifts of 7.5 working hours each in a day. Only one shift was considered for study, so the total time in seconds comes out to be:

$$\text{Time per shift} = 7.5 \times 60 \times 60 \text{ seconds}$$

Thus, according to the established cycle time of 4.0 seconds, the production target should be:

$$\text{Target} = (7.5 \times 60 \times 60)/4 = 6750$$

But since the line runs on variable cycle times in actuality, this production target becomes hard to achieve. Thus, the system also shows the projected target based on actual cycle time on a real-time basis.

The established cycle time of the assembly line, where system was tested, was 4.0 seconds. So, this was taken as the standard cycle time on all three workstations. Whenever the actual cycle time exceeded the established cycle time, the system showed a warning as a delay.

In Figure 11.11, images shown are of the actual results from the testing. Serial monitor of Arduino IDE was used for display purposes but other channels can also be used like specially designed HMIs, etc.

11.5 DISCUSSION

In the three-hour testing on each workstation, the results can be summarized as:

1. The bottleneck changed from one station to another several times during the testing.

TABLE 11.2
Test Summary

Workstation	Nucleus Caulking	Resonator Riveting	Air Gap Checking
Cycle Time (sec)	4.0	4.0	4.0
Run time for Test	3 hours	3 hours	3 hours
Frequency of Cycle Time Breached	157	241	163

2. Out of three workstations, resonator riveting was the bottleneck most of the time.
3. The cycle time breaches during the three-hour testing are shown in Table 11.2.

The results obtained after testing the system in the industry show that the difference in the actual cycle times of the same workstation is not minor in numbers. There is a significant difference between the actual cycle times of the same workstation.

Also, it is clear from the results that delay on a workstation is occurring several times, which indicates the inefficiency of the assembly line. This inefficiency is the result of either the inaccuracy in the work done by the operator or any other reason.

The projected target of the day is an important parameter to evaluate. The results show that in some instances, the projected target comes out to be less than the production target, which is not in favor of the industry. So, one can take corrective measures and improve the projected target.

The motive of the project was to develop a monitoring system to warn of the problems and enable the managers to directly look at the places of origin of these problems instead of searching the whole line. However, taking corrective measures depends upon the capability and willpower of managers. Thus, effectively using the system is the sole responsibility of the company.

11.6 LIMITATIONS AND CHALLENGES OF PROJECT

The project has its own limitations while implementing:

1. The system was applied on three workstations only and this can be extended to the whole assembly line.
2. The system is not reliable for harsh industrial environments.
3. Arduino was preferred in the project because of the small assembly lines. In large industries, Arduino can be replaced by PLCs.

11.7 FUTURE SCOPE

Extending this concept for factory-wide application can be done for future endeavors. Other industries can be included for this project implantation, which may include a food processing unit, pharmaceutical manufacturing line, etc., for monitoring on a real-time basis. As the system collected the necessary data, it can also calculate other crucial measures like balance delay, line efficiency, smoothness index, etc. The system can display the data on an excel spreadsheet using third-party communication software like RSLINX, etc. This will result in easy handling of data.

The biggest revolution can be brought by integrating machine learning with this system so that the system will itself take corrective measures whenever required. The system will contain the previous record of data and will self-analyze it to make the decision when there is an indication of delay on any workstation. Thus, applications of data sciences can have great implications when integrated with this system.

Ranking of the efficient workers with the help of recorded data can be introduced as a work-study project using this technology. So, this will help to select the best workers so that the cycle time can be reduced further in future.

PLCs together with Ladder logic programming, Relays, etc., can be a good substitute for microcontrollers for large industries. Thus, the system has a wide scope for future study and extension.

11.8 CONCLUSION

This project was able to identify the number of cycle time breaches and thus the actual bottleneck workstation which varies with time. Various measures like the actual cycle time, projected target of the day, and indication of the delay on the workstation were easily detected by the system proposed in this project. As the system was installed only on three workstations for the purpose of testing, the real-time monitoring and analysis of the complete assembly line were not possible. Although the results obtained show that the system can perform real-time monitoring and analysis of the complete assembly line making it an easy task.

Few of the benefits of employing this system are:

1. Actual cycle time of each workstation can be found on a real-time basis.
2. Identification of bottlenecks on a real-time basis.
3. The system will keep a record of the number of cycle time breaches on each workstation.
4. The system can identify inefficient workers by recording the cycle time history of each workstation for each worker.
5. Effective measures can be taken to remove bottlenecks at the same time when the system gives the information. This can prevent underproduction.
6. It lacks the ruggedness required for the industrial application. Also, the system cannot be used for bigger projects.

ACKNOWLEDGMENT

The authors would like to thank Minda Industries Limited, Rudrapur, for their support during this work.

REFERENCES

Abuhasel, K.A. and Khan, M.A., 2020. A secure industrial Internet of Things (IIoT) framework for resource management in smart manufacturing. IEEE Access, 8, p. 117354–117364.

Borra, P. and Shinde, Y., 2017. Incorporation of IoT in Assembly Line Monitoring System. International Research Journal of Engineering and Technology, 4(12), p. 1691–1693.

Calderón Godoy, A.J. and González Pérez, I., 2018. Integration of sensor and actuator networks and the SCADA system to promote the migration of the legacy flexible manufacturing system towards the industry 4.0 concept. Journal of Sensor and Actuator Networks, 7(2), p. 23.

Cheung, W.F., Lin, T.H. and Lin, Y.C., 2018. A real-time construction safety monitoring system for hazardous gas integrating wireless sensor network and building information modeling technologies. Sensors, 18(2), p. 436.

Ebling, M.R. and Want, R., 2017. Pervasive computing revisited. IEEE Pervasive Computing, 16(3), p. 17–19.

Gubbi, J., Buyya, R., Marusic, S. and Palaniswami, M., 2013. Internet of Things (IoT): A vision, architectural elements, and future directions. Future Generation Computer Systems, 29(7), p. 1645–1660.

Hou, W., Guo, L. and Ning, Z., 2019. Local electricity storage for blockchain-based energy trading in industrial internet of things. IEEE Transactions on Industrial Informatics, 15(6), p. 3610–3619.

Khan, M.A. and Abuhasel, K.A., 2020. Advanced metameric dimension framework for heterogeneous industrial Internet of things. Computational Intelligence, p. 1–21.

Khan, M.A. and Alghamdi, N.S., 2021. A neutrosophic WPM-based machine learning model for device trust in industrial internet of things. Journal of Ambient Intelligence and Humanized Computing. https://doi.org/10.1007/s12652-021-03431-2

Kortuem, G., Kawsar, F., Sundramoorthy, V. and Fitton, D., 2009. Smart objects as building blocks for the internet of things. IEEE Internet Computing, 14(1), p. 44–51.

Kumar, N. and Mahto, D., 2013. Assembly line balancing: a review of developments and trends in approach to industrial application. Global Journal of Research in Engineering, 13(2), p. 29–50.

Lee, J., Noh, S.D., Kim, H.J. and Kang, Y.S., 2018. Implementation of cyber-physical production systems for quality prediction and operation control in metal casting. Sensors, 18(5), p. 1428.

Li, J., Xie, J., Yang, Z. and Li, J., 2018. Fault diagnosis method for a mine hoist in the internet of things environment. Sensors, 18(6), p. 1920.

Manes, G., Collodi, G., Gelpi, L., Fusco, R., Ricci, G., Manes, A. and Passafiume, M., 2016. Realtime gas emission monitoring at hazardous sites using a distributed point-source sensing infrastructure. Sensors, 16(1), p. 121.

Moon, Y.S., Choi, H.R., Kim, J.J., Kim, D.W., Cho, J.H., Kim, J.W. and Jeong, J.W., 2017. Development of IoT-based sensor tag for smart factory. International Research Journal of Electronics & Computer Engineering, 3, p. 28–31.

Roshani, A. and Giglio, D., 2017. Simulated annealing algorithms for the multi-manned assembly line balancing problem: minimising cycle time. International Journal of Production Research, 55(10), p. 2731–2751.

Ruppert, T. and Abonyi, J., 2018. Industrial internet of things-based cycle time control of assembly lines. In 2018 IEEE International Conference on Future IoT Technologies, p. 1–4.

Salamone, F., Danza, L., Meroni, I. and Pollastro, M.C., 2017. A low-cost environmental monitoring system: how to prevent systematic errors in the design phase through the combined use of additive manufacturing and thermographic techniques. Sensors, 17(4), p. 828.

Shukla, P., Malviya, S. and Jain, S., 2018. Review of some recent findings for productivity improvement using line balancing heuristic algorithms. International Journal of Innovative Research in Technology, 5(6), p. 83–90.

Subramaniam, S., Husin, S., Yusop, Y. and Hamidon A., 2007. Real time shop floor monitoring system for a better production line management. Proceeding of Asia-Pacific Conference on Applied Electromagnetics, p. 1–4.

Syafrudin, M., Alfian, G., Fitriyani, N.L. and Rhee, J., 2018. Performance analysis of IoT-based sensor, big data processing, and machine learning model for real-time monitoring system in automotive manufacturing. Sensors, 18(9), p. 2946.

Syafrudin, M., Lee, K., Fitriyani, N.L., Alfian, G., and Rhee, J., 2017. Real time monitoring system based on IoT sensor data for assembly line process in the automotive factory. Proceedings of the Korean Society for Production and Manufacturing Conference, p. 200–200.

Taifa, I. and Vhora, T., 2019. Cycle time reduction for productivity improvement in the manufacturing industry. Journal of Industrial Engineering and Management Studies, 6(2), p. 147–164.

Xu, Y. and Chen, M., 2016. Improving just-in-time manufacturing operations by using Internet of Things based solutions. Procedia CIRP, 56, p. 326–331.

Xu, Y. and Chen, M., 2018. An Internet of Things based framework to enhance just-in-time manufacturing. Proceedings of the Institution of Mechanical Engineers, Part B: Journal of Engineering Manufacture, 232(13), p. 2353–2363.

Zait, A., 2018. An introduction to arduino uno pinout. https://www.circuito.io/blog/arduino-uno-pinout.amp/

Zhang, X., Zhang, J., Li, L., Zhang, Y. and Yang, G., 2017. Monitoring citrus soil moisture and nutrients using an IoT based system. Sensors, 17(3), p. 447.

12 IoT-Enabled Hazardous Gas Leakage Detection System for Citizen's Safety

Prerna Sharma and Latika Kharb
Jagan Institute of Management Studies (JIMS)
Delhi, India

CONTENTS

12.1 Introduction 257
12.2 Literature Review 259
12.3 System Architecture 260
    12.3.1 Arduino and ATMEGA328P..260
    12.3.2 MQ2 Sensor Unit 261
12.4 Hardware Assembly and Implementation 261
    12.4.1 MQ2 Sensor Integration with an Arduino Board 261
    12.4.2 Connections 263
    12.4.3 MQ2 Sensor Integration Code Snippet..263
    12.4.4 Working 263
    12.4.5 Applications 264
12.5 Future Scope 266
12.6 Conclusion 266
Appendix..267
References..268

12.1 INTRODUCTION

Internet of Things (IoT) explains that any device can connect and communicate with some other devices over the network. With electronic devices such as sensors and actuators, the network's connected objects could share the needful data with other devices to accomplish the task. The IoT is based on four blocks, namely:

- ***Sensors:*** Sensors can detect a wide range of information from the atmosphere and location transferred to the IoT gateway framework. For example, temperature sensor will collect the temperature of the room or place at a particular time and transmit it to some other objects.

- **_IoT framework and gateway:_** From the name itself, quickly know that the gateway acts as a path between the internal objects to the external things connected through the Internet. With the help of a gateway, only the internal objects can share the information at the right time. The IoT framework and gateway collect the data from the sensors and transmit it to the connected external things over the Internet.
- **_Cloud server:_** The cloud server collects the transmitted data from the gateway and stores it to the cloud server for further process. The stored data helps to perform intelligent activities to efficiently fulfill user requirements and make all the connected devices a smart device. The cloud server is responsible for the decision-making against the challenges faced in real time by the end user.
- **_Mobile application:_** Mobile application is another essential tool used to control the user's whole system anywhere and anytime. The user can see the cloud server's analyzed information as a dataset or any diagrammatic representation such as graphs and images.

Hazardous gas leakage detection is one of the primary aspects of scrutinizing a place's security and a critical element in the security system both at the household and the industry levels. Apart from security purposes, this very same approach is applicable as the standard for detecting the target compound in various processes of factories, industries, and underground mines as the core functionality. Moreover, due to the urbanization and material commercialization of various toxic compounds–producing sectors, it has become the need of the hour to keep a check on these parameters to a notable degree. Also, the consideration of smoke and pollutants can't go unwitnessed while discussing the air quality. Hence, the microcontroller and sensor-based approach could also be applied here for monitoring and security purposes. As a result, there is a growing demand for air quality monitoring and control systems. It is noteworthy that all the areas mentioned earlier could be potential applications of the prototype and algorithm discussed in this chapter, while on lower grounds using slightly different hardware components to achieve desired and more specific solutions. The organizations and authorities are still using the same traditional approaches when it comes to security solutions to standard target compounds and leakage detections. Advancements in the sensor technology have brought a significant change in this scenario. As an inexpensive and effective solution to compound detection, the advent of the MQ sensor series resolves a need for a standard solution. MQ sensors are metal oxide-based (MOS)-type sensors substantially employed for the detection of explosive steam and flammable gases. MOS-type sensors vary their electrical resistance with respect to concentration of target gases in the vicinity and can be used to sample the air nearby to detect the presence of target compound. Each unit in the MQ family is susceptible to a different set of multiple compounds, making the series practical and employable for a wide variety of applications while being a cost-effective standard. Integrating numerous apparatuses and mechanisms makes up the system prototype's design into the system. The concerned personnel get an alarm when the MQ sensor smells a hazardous compound and takes automated actions to prevent the potential damage to the model environment. However, this chapter focuses on the elementary aspect of achieving the actuation mechanism of alarming the authorities by integrating an alarm/buzzer (a quick reference to these channels and procedures to incorporate more complex actuation mechanisms through this chapter). This chapter proposes a system capable of detecting hazardous gas leakage and smoke and alarming the concerned authorities. The expected outcome of the plan is to operate without any human intervention. The compound detection aspect of the proposed system proves to be a piece of low-powered equipment; however, the power consumed by the actuation aspect of the system varies and solely depends on the energy consumed by the interfaced components as per requirements. The system's development consists of three segments: target gas detection, decision-making and operativeness, and alarming (actuation mechanism) the concerned authorities. The detailed information about the system design follows in the further sections of the chapter.

12.2 LITERATURE REVIEW

In today's scenario, it seems pretty apparent that there are multiple and distinguished ways to approach, comprehend, and solve a problem, especially technological ones. In other words, we can say that there exist numerous algorithms and several strategies to get started with and approach the issue; the same goes with the modules to achieve the gas detection feature. Researchers and engineers in their respective fields have developed several strategies to achieve the same throughout this time. Some of the viable solutions, including both the traditional and new innovative ones, are mentioned here. Luay Fraiwan et al. (2011) developed a wireless gas leakage detection system for households, foundries, and business units depending on the use of LPG and natural gas in their processes. The proposed architecture senses the variance in the concentration of target gases with respect to a set threshold value; it activates an audiovisual alarm and triggers an alarming actuator provided with mobility to alarm the people in the premises.

Amin et al. (2018) featured a model of a cigarette disposal blower and automatic freshener using an MQ-5 sensor based on an AVR Atmega 8535 microcontroller. The model built in the process can blow the smoke and automatically fill the room with fragrance, prevent the build-up of cigarette smoke in a room, and help minimize uneasiness for passive smokers so that the air circulation in the room space continues to be maintained. The system employs the MQ-135 unit to detect smoke and target gases. The device is connected to a remote server and continuously updates the sampling results on the webserver. Hence, real-time results are available on-site via an LCD display interfaced with the system and anyone to access remotely. Thomas G. McRae, Jr (n.d.) invented a video imaging system for detecting hazardous gas leakages. The author has discussed his algorithm as a backscatter absorption gas imaging system. Abdulrafy et al. ((n.d.) developed a prototype of a gas detector using the Arduino prototyping board targeted for laboratories and factories, that is, majorly for industry standards. The system aims explicitly to target natural gas and LPG detection. The prototype and detection also tend to reroute the target gas and smoke using an exhaust fan and exchange messages with the remote authorities in case of potential threat. Jadin and Ghazali et al. (2014) proposed a gas leakage detection strategy using a thermal (infrared image analysis) imaging technique. Amanda W. Lewis et al. (2003) proposed a gas leakage methodology using infrared thermography. The author also discussed various other methods to detect landfill gas emissions and leakage regarding fundamental factors, such as weather conditions, distance from the sensor itself, and ground conditions. Varma et al. (2017) proposed a gas leakage detection model with real-time data monitoring and possible threat analysis by data analysis. The developed model provides the feature of sending a text message and an email and making a call when there is a leakage threat (or just the possibility of the same).

12.3 SYSTEM ARCHITECTURE

12.3.1 Arduino and ATMEGA328P

It is a microcontroller from the MegaAVR series. It is a low-power, CMOS 8-bit microcontroller based on the AVR-enhanced reduced instruction set computer (RISC) architecture with high endurance, non-volatile memory segments as discussed in ATMEL (n.d.) and Microchip (n.d.). Figure 12.1 exhibits the pin-out structure of ATMEGA328P. We wrote the programs in C++ defining the system's functionality following all the particular rules of code structuring, also known as sketches in the Arduino integrated development environment as discussed in Sharma and Kamthania (2020). We deployed these sketches on this chip to define the operations of the system. It acts as the core of the system.

This chip also comes embedded in Arduinos. Arduino is an open-source electronics platform based on easy-to-use hardware and software; it senses the environment by receiving inputs from many sensors and affects its surroundings by controlling lights, motors, and other actuators as in Arduino (n.d.). Arduino board along with software developing kit (SDK)-based programming is used for interacting and commanding different sensors as in Singh et al. (2021).

Arduinos is a very inexpensive and easy solution to microcontrollers; using the chip individually could be cumbersome but with Arduino, programs can be uploaded via USB very easily and no previous knowledge and experience with microcontrollers are required as discussed in Sharma and Kamthania (2019).

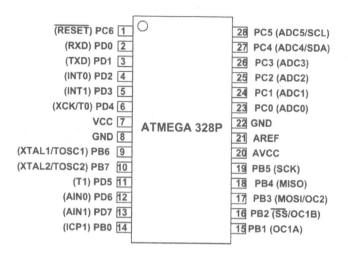

FIGURE 12.1 Atmega328P pinout. (Atmega328 DataSheet.)

Arduino also offers various features like serial monitor debugging tools, customizable script development environment, accessibility/setting multiple environment variables for deployment of sketches, and serial data communication transfer rate. The other aspects of the system include interfacing the sensors and actuators with Arduino via various digital and analog pins.

The development and manufacturing of the leakage detection system proposed in this chapter revolve around the Arduino prototyping environment. The deployed module works for the decision-making process. So, the research work focuses on incorporating technology into people's life without disturbing their daily routine and does not require separate time to use as discussed in Singh et al. (2020). It will take necessary action through actuators once it gets notified about the unusual target gas concentration pattern through the attached sensor(s): the code structure and the system's working is shown in disclosure format in Part IV.

12.3.2 MQ2 Sensor Unit

MQ2 module is a MOS-type sensor from the MQ sensor series, a widely adopted standard for target compound detection. MOS sensors are often termed as chemically resistive modules as their electrical resistance tends to vary following the concentration of target gases in their vicinity. Every sensor in the MQ sensor series is sensitive to a different set of multiple gases, making the MQ sensor series apt, robust, and cost-effective in standard target gases and steam detection. The MQ2 sensor unit can detect methane, butane, propane, hydrogen, LPG, and smoke as discussed in MQ-2 (n.d.). MQ6 (targeting gases such as LPG, butane, and methane), MQ306A (with a high sensitivity toward LPG and butane), and MQ309A (targeting gases such as carbon monoxide and methane) can be used as alternatives to MQ2. But MQ2 is the best choice for detecting the current scenario and requirements as both include combustible gases and smoke. MQ2 sensor operates with a working voltage typically at 5V (low power consumption) with a preheat duration of approximately less than 20 seconds (The requirement includes a sensor to heat up to start the sampling procedure. Each time the system gets powered on, it will take some time before operating; however, there will be no time overheads in the corresponding loop cycles of the system.) The proposed approach in this chapter aims to identify the suspension and relative concentration of only methane, butane, propane, hydrogen, LPG, and smoke in the model environment. Therefore, as we advance with the MQ2 sensor, it is noteworthy that any sensor(s) from the MQ series can be selected according to the requirements and integrated with Arduino.

12.4 HARDWARE ASSEMBLY AND IMPLEMENTATION

12.4.1 MQ2 Sensor Integration with an Arduino Board

Arduino is used as a prototyping board and offers multiple advantages in deploying and debugging external plugins, thus providing easy, quick, and reliable integration of various sensors and actuators. As shown in Figure 12.2, Arduino Uno has 13 digital and 6 analog input/output pins with the pulse width modulation (PWM) functionality available on only 6 pins (3, 5, 6, 9, 10, and 11).

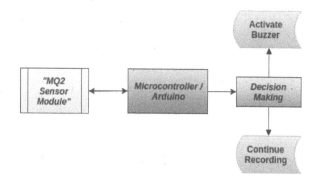

FIGURE 12.2 System architecture.

MQ2 sensor needs to be interfacing with the Arduino as a sampling component. Both analog and digital I/O pins of the Arduino interface the sensor since MQ2 can function as digital or analog sensors. MQ2 unit comes with six pins (2H, 2A, and 2B) as shown in Figure 12.3, whereas the sensor module (packaged unit) has only four pins (Vcc, GND, digital output, and analog output) as displayed in Figure 12.4.

FIGURE 12.3 MQ2 unit.

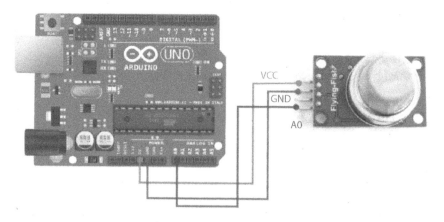

FIGURE 12.4 Arduino interfacing with MQ2 circuit diagram.

12.4.2 Connections

- GND to GND pin of Arduino
- Vcc to 5V pin of Arduino
- AO (analog output) to A0 pin of Arduino
- DO (digital output) to none

12.4.3 MQ2 Sensor Integration Code Snippet

```
int buzzer = 13;
int smokeA0 = A5;

// Your threshold value. You might need to change it.
int sensorThreshold = 500;

void setup() {
  pinMode(buzzer, OUTPUT);
  pinMode(smokeA0, INPUT);
  Serial.begin(9600);
}

void loop() {

int analogSensor = analogRead(smokeA0);

  Serial.print("Pin A0: ");
  Serial.println(analogSensor);
  // Checks if it has reached the threshold value

  delay(1000);

  if(analogSensor>sensorThres)
  {
      digitalWrite(13,HIGH);
      Serial.println("High");
  }
  else
  {
      digitalWrite(13,LOW);
      Serial.println("Low");
  }
}
```

12.4.4 Working

All the interfacing components with Arduino follow the procedure as discussed here. Once the system is powered, it will implicitly prompt the sensor, that is, MQ2, to prepare for the sampling procedure. The sensor module will prepare to record the data. Two out of six pins of the sensor dedicated for heating gets activated as a result of this. It will take approximately around 20–30 seconds to preheat to start recording data. Once initialization starts with heating, the recorded values gradually begin to represent the actual concentration, unlike a random set of values produced during the preheating phase. It will then start its sampling procedures.

It is programed to continuously sample the air in its vicinity in the reserved space after a periodic interval of 1000 ms. The sensor gets calibrated according to the model environment of testing with 400 units, which depicts the threshold range. Every sensor unit needs to be tested in the model environment to obtain this threshold value; in our case, it is 400.

The sensor is constantly updating the relative concentration of the target compound and the microcontroller compared to those values with the threshold range. In case it detects the presence of a target gas (disclosed above) in a higher concentration for the preset threshold value, it gets notified. The microcontroller will then generate a low output to the digital pin to which the actuator gets interfaced to trigger the same; in our case, it is a buzzer.

Meanwhile, for the entire duration, this buzzer is activated, the sensor will continue recording and updating values. If the current value comes under the valid range of threshold bounds, the alarm gets called off; else, it will continue to buzz. This alert mechanism will alert the concerned authorities about the event so that the admin takes the required actions as soon as possible, consequently avoiding the damage. As discussed above, any complex mechanism gets integrated with the system as an actuator through the same channel. The microcontroller will activate those channels to take action. Flowchart for the gas leakage detection strategy is disclosed in Figure 12.5.

12.4.5 APPLICATIONS

In this chapter, a robust gas leakage detection and monitoring system is developed. This model is just a fundamental step toward a bigger goal. The proposed system is capable of detecting target flammable gases and combustible steam and therefore alarming the concerned authorities about the outliers and thus preventing the potential threat. As this model is built as a standard solution for monitoring the potential risk of damage due to gas leakage, it can be substantially employed wherever there's a risk of gas leakage. It is installed in offices, educational institutes, hospitals, mines, factories, foundries, industrial plants, and residents. The system is like a small container installed without any significant change in the model environment. All the hardware used in the development of the system is widely available, low-powered, and inexpensive, making the device affordable and maintainable. Availability and accessibility of this kind of device can adjourn many disastrous situations like the Westray Coal Mine disaster (Canada, May 9, 1992) and Bhopal Gas Tragedy (India, December 2, 1984). The system can also be used as a prototype for students and researchers to deploy algorithms and data mining techniques to avoid potential risk.

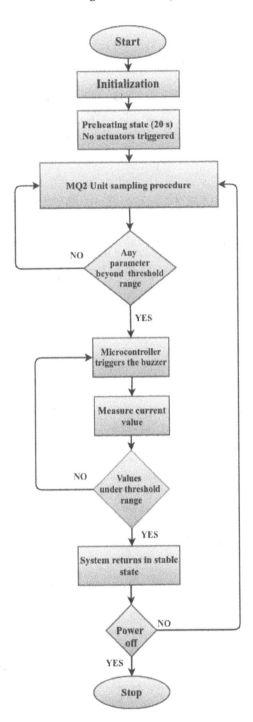

FIGURE 12.5 Flowchart of gas leakage detection strategy.

12.5 FUTURE SCOPE

In the future, the system will be more compact with expanded sensor technology (customized set of the sensor(s) from the MQ series), including the capabilities and mechanism to monitor the gas concentration remotely. An additional Wi-Fi chip can be integrated into the system to establish network communication with remote servers. The server will store all logs of the sampling procedure enabling the user to monitor and recover older records. This feature of data persistence could serve as a repository for the task of research and analysis, detection of potential risk, and redefining the threshold values. The owner or the concerned individual can track the conditions of a reserved area and get notified in case there is a detection of gas leakage. This approach could be beneficial in a locale like mines and refineries.

12.6 CONCLUSION

IoT has many applications in the area of communication, edge computing, healthcare, and industrial processes (Abuhasel and Khan, 2020; Alghamdi and Khan, 2021; Khan and Alghamdi, 2021; Khan et al., 2020; Mukherjee et al., 2021; Munusamy et al., 2021; Nandy et al., 2021; Quasim et al., 2021; Verma et al., 2021; Xu et al., 2021). The goal of this chapter was to develop a robust hazardous gas leakage detection and alert system. The system is built on the Arduino platform with MOS-based sensing technology to detect the presence of dangerous gas or smoke, employing a low-power buzzer for the alert. The developed system was tested freely in LPG gas and smoke and produced desirable results. MQ2 unit was used for sampling the air in the vicinity and record the relative concentration of target compounds. MQ2 unit is susceptible to methane, butane, propane, hydrogen, LPG, and smoke. The system becomes stable to record the data in approximately 20 seconds since the MQ2 sensor gets preheated for 20 seconds. One can use Arduino for local intelligence. The sketch defining the instructions gets deployed on the Arduino through USB. Once initiated, it continuously samples the model environment and checks against the threshold values; if the recorded values are out of the legal range (more significant than the threshold), it activates the actuators, a buzzer in our case to alert the concerned authorities and personnel about the leakage. It can be recorded as an observation that it reacted with smoke a little earlier than LPG. MQ sensor series offers multiple modules with a different set of gases that a particular unit can detect. Hence, we interface the desired unit according to the model environment's requirements. Alternative techniques to see a specific compound other than sensor technology could be thermal imaging, infrared cameras, the employment of polar chemical compounds, etc. The developed system can be installed anywhere where security of people and facility is involved, with minor alterations in the setup. The system will operate without any human intervention.

APPENDIX

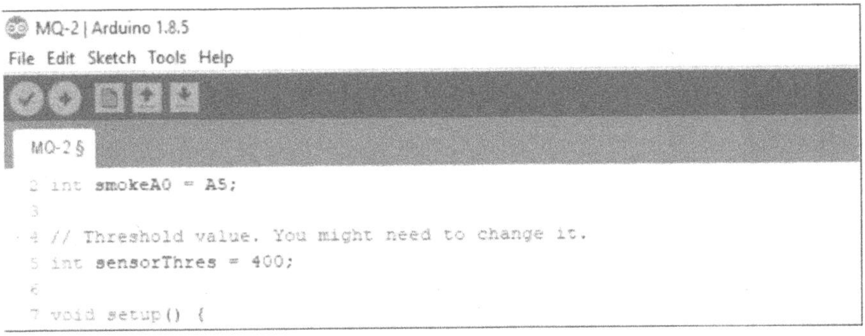

```
MQ-2 | Arduino 1.8.5
File Edit Sketch Tools Help

2  int smokeA0 = A5;
3
4  // Threshold value. You might need to change it.
5  int sensorThres = 400;
6
7  void setup() {
```

FIGURE 12.6 Threshold value being set for the decision-making process.

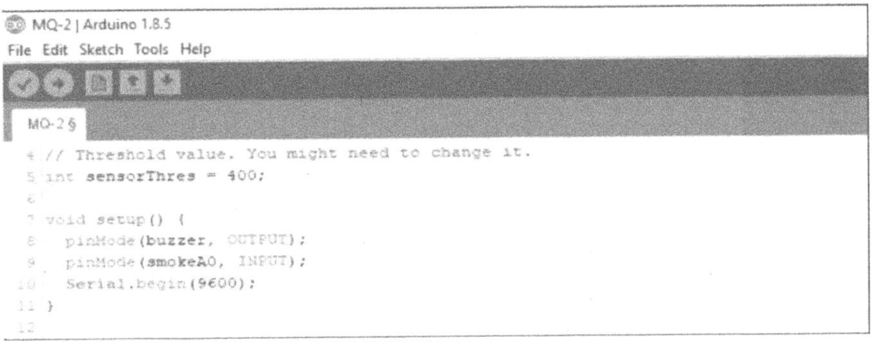

```
MQ-2 | Arduino 1.8.5
File Edit Sketch Tools Help

4  // Threshold value. You might need to change it.
5  int sensorThres = 400;
6
7  void setup() {
8    pinMode(buzzer, OUTPUT);
9    pinMode(smokeA0, INPUT);
10   Serial.begin(9600);
11 }
```

FIGURE 12.7 MQ2 Module being set up as input of the system.

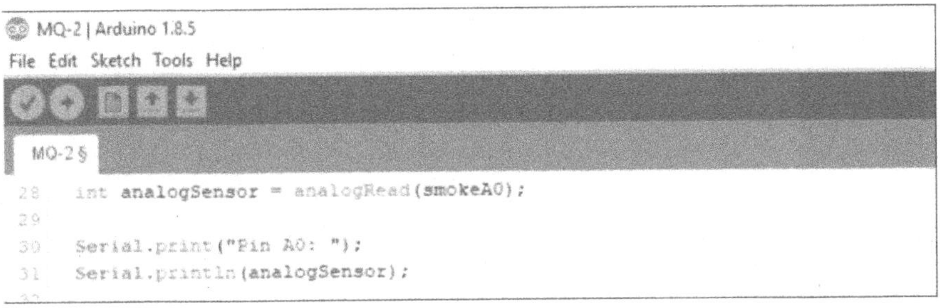

```
MQ-2 | Arduino 1.8.5
File Edit Sketch Tools Help

28  int analogSensor = analogRead(smokeA0);
29
30  Serial.print("Pin A0: ");
31  Serial.println(analogSensor);
```

FIGURE 12.8 Reading Sensor value using built-in function -analog Read ().

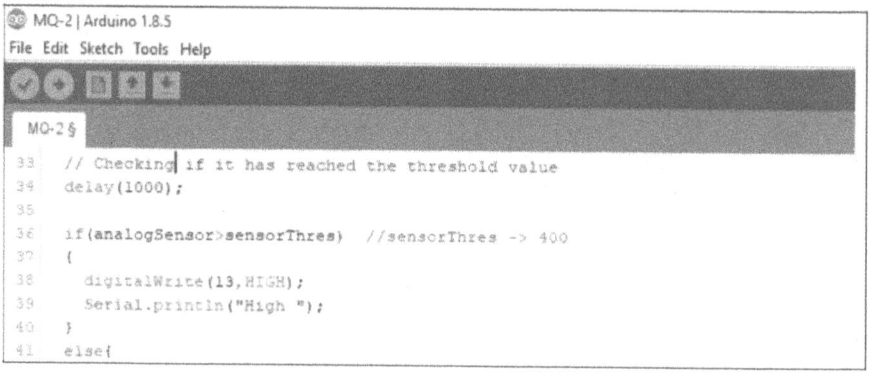

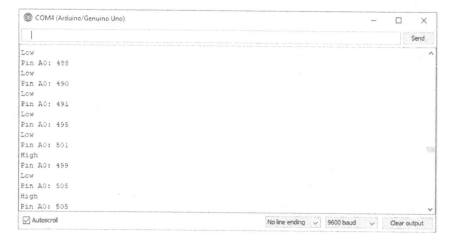

FIGURE 12.9 Decision making on the basis of the threshold value.

FIGURE 12.10 Sensor calibration testing at *threshold value: 500*.

REFERENCES

Abdulrafy, G., Omar, Ameer, Yasin, A., Ditual, Jheryll E., Urot, Mohammad Reza, Dimal, G., Nahid, G., Mamco, and Sagarino, Chris T. (n.d.), "MQ2-TECTOR: An Arduino Based Gas Detector, Preventing Gas-leak Explosion," doi: 10.13140/RG.2.2.15517.56808.

Abuhasel, K. A. and Khan, M. A. "A Secure Industrial Internet of Things (IIoT) Framework for Resource Management in Smart Manufacturing," IEEE Access, vol. 8, pp. 117354–117364, 2020. doi: 10.1109/ACCESS.2020.3004711.

Alghamdi, N. S. and Khan, M. A. "Energy-Efficient and Blockchain-Enabled Model for Internet of Things (IoT) in Smart Cities," Computers, Materials & Continua, vol. 66, no. 3, pp. 2509–2524, 2021.

Amin, M. M., Azel Aji Nugratama, M., Maseleno, A., Huda, M., and Azmi Jasmi, K. "Design of Cigarette Disposal Blower and Automatic Freshener Using MQ-5 Sensor Based on ATMEGA 8535 Microcontroller," International Journal of Engineering & Technology, vol. 7, pp. 1108–1113, 2018. doi: 10.14419/ijet.v7i3.11917.

Arduino. (n.d.). "Arduino.cc.Official Arduino Documentation" [Online]. Retrieved from https://www.arduino.cc/en/Guide/Introduction.

ATMEL. (n.d.). ATMEGA328P datasheet. ATMEL 8025I-AVR-02/09

Fraiwan, L., Lweesy, K., Bani-Salma, A., and Mani, N. "A Wireless Home Safety Gas Leakage Detection System." 1st Middle East Conference on Biomedical Engineering, Sharjah, United Arab Emirates: IEEE, pp. 11–14, 2011. doi: 10.1109/MECBME.2011.5752053.

Jadin, M. S., and Ghazali, K. H. "Gas Leakage Detection Using Thermal Imaging Technique." 2014 UKSim-AMSS 16th International Conference on Computer Modelling and Simulation, pp. 302–306, 2014.

Khan, M. A., and Alghamdi, N. S. A Neutrosophic WPM-Based Machine Learning Model for Device Trust in Industrial Internet of Things. Journal of Ambient Intelligence and Humanized Computing. 2021. doi: 10.1007/s12652-021-03431-2.

Khan, M. A., Quasim, M. T., Alghamdi, N. S., and Khan, M. Y. "A Secure Framework for Authentication and Encryption Using Improved ECC for IoT-Based Medical Sensor Data," IEEE Access, vol. 8, pp. 52018–52027, 2020. doi: 10.1109/ACCESS.2020.2980739.

Lewis, A. W., Yuen, S. T. S., and Smith, A. J. R. "Detection of Gas Leakage from Landfills Using Infrared Thermography—Applicability and Limitations," Waste Management & Research, vol. 21, no. 5, pp. 436–447, 2003. doi: 10.1177/0734242X0302100506.

McRae, Thomas G., Jr. (n.d.), "Backscatter absorption gas imaging system" Patent US4555627A.

Microchip. (n.d.). "Microchip" [Online]. ATMEGA328P. Retrieved from https://www.microchip.com/wwwproducts/en/ATmega328

MQ-2. (n.d.). "'MQ-2 Datasheet' MQ-2 Semiconductor Sensor for Combustible Gas" [Online]. Retrieved from https://www.pololu.com/file/0J309/MQ2.pdf

Mukherjee, A., Goswami, P., Khan, M. A., Manman, L., Yang, L., and Pillai, P. "Energy-Efficient Resource Allocation Strategy in Massive IoT for Industrial 6G Applications," IEEE Internet of Things Journal, vol. 8, no. 7, pp. 5194–5201, April 1, 2021. doi: 10.1109/JIOT.2020.3035608.

Munusamy, A., et al. "Service Deployment Strategy for Predictive Analysis of FinTech IoT Applications in Edge Networks," IEEE Internet of Things Journal. 2021. doi: 10.1109/JIOT.2021.3078148.

Nandy, S., Adhikari, M., Khan, M. A., Menon, V. G., and Verma, S. "An Intrusion Detection Mechanism for Secured IoMT framework Based on Swarm-Neural Network," IEEE Journal of Biomedical and Health Informatics, pp. 1–1, 2021. doi: 10.1109/JBHI.2021.3101686.

Quasim, M. T., et al. "Emotion-Based Music Recommendation and Classification Using Machine Learning with IoT Framework," Soft Computing, vol. 25, 12249–12260, 2021. doi: org/10.1007/s00500-021-05898-9.

Sharma, P., and Kamthania, D. (2019). "Intelligent Object Detection and Avoidance System," International Conference on Transforming Ideas (Inter-Disciplinary Exchanges, Analysis, and Search) into Viable Solutions. Macmillan Education. ISBN-938882695-7.

Sharma, P., and Kamthania, D. (2020), "Intrusion Detection and Security System," Tanwar, P., Jain, V., Liu, C.-M. and Goyal, V. (eds), Big Data Analytics and Intelligence: A Perspective for Health Care. Emerald Publishing Limited, Bingley, pp. 139–151. doi: org/10.1108/978-1-83909-099-820201011.

Singh, R., Singh P., Kharb L. (2020), "Proposing Real-Time Smart Healthcare Model Using IoT," Raj, P., Chatterjee, J., Kumar, A., and Balamurugan, B. (eds), Internet of Things Use Cases for the Healthcare Industry. Springer, Cham, pp. 949–964. doi: org/10.1007/978-3-030-37526-3_2.

Singh R., Singh, P., Chahal, D., and Kharb, L. (2021), "'VISIO': An IoT Device for Assistance of Visually Challenged," Pandey, V. C., Pandey, P. M., Garg, S. K. (eds), Advances in Electromechanical Technologies. Lecture Notes in Mechanical Engineering. Springer, Singapore. doi: 10.1007/978-981-15-5463-6_84.

Varma, A., Prabhakar, S., and Jayavel, K. "Gas Leakage Detection and Smart Alerting and prediction using IoT," 2nd International Conference on Computing and Communications Technologies (ICCCT), Chennai, India, pp. 327–333, 2017. doi: 10.1109/ICCCT2.2017.7972304.

Verma, S., Kaur, S., Khan, M. A., and Sehdev, P. S., "Toward Green Communication in 6G-Enabled Massive Internet of Things," IEEE Internet of Things Journal, vol. 8, no. 7, pp. 5408–5415, 2021. doi: 10.1109/JIOT.2020.3038804.

Xu, L., Zhou, X., Khan, M. A., Li, X., Menon, V. G., and Yu, X. "Communication Quality Prediction for Internet of Vehicle (IoV) Networks: An Elman Approach," IEEE Transactions on Intelligent Transportation Systems. 2021. doi: 10.1109/TITS.2021.3088862.

Index

A

Active Reader Active Tag (ARAT), 59
Active Reader Passive Tag (ARPT), 59
Actuators, 28, 136
Ad-hoc on-demand distance vector reactive
    protocols (AODV), 210
Advanced message queuing protocol (AMQP),
    60, 156
Application layer, 40, 45, 151
Assembly line balancing (ALB), 240

B

Bluetooth, 58, 148, 170
Bluetooth Low Energy (BLE), 148
Body Area Network (BAN), 62, 64
Body Sensor Network (BSN), 64
Building management systems (BMS), 16
Bus, 15, 77

C

Cache memory architecture (CMA), 80, 90
Central management system (CMS), 231
Cloud computing layer, 49
Cloud server, 258
CMOS TTL, 21
Communication protocol, 19, 56, 155, 231
Complementary metal-oxide-semiconductor
    (CMOS), 76
Complexity, 7, 27, 57, 64
Constrained application protocol (CoAP), 60, 156
Convolutional neural network (CNN), 111, 120

D

Data analysis, 47
Data concentrator unit (DCU), 231
Data ingestion, 47, 234
Deep neural network (DNN), 108
Differential sense amplifier, 83, 84, 85
Direct-sequence spread spectrum (DSSS), 145
Distributed denial of service (DDoS), 175
Dynamic random-access memory (DRAM), 78
Dynamic source control routing (DSR), 210

E

Edge computing, 48
EnOcean, 153

F

Fog computing layer, 49
FPGA, 72, 114, 123

G

Gateway (GW), 16, 18, 30
Global positioning system (GPS), 209

H

Heterogeneity, 5, 108, 162, 171, 185
Hybrid, 17, 36, 48

I

Information examination, 47
Intel, 4
Inter IC communication (I2C), 76
International Technology Roadmap for
    Semiconductors (ITRS), 78
Internet of Things (IoT), 3, 39, 43, 56, 76,
    135, 258
Internet of Vehicle (IoV), 205

L

LM35, 34
Load balancing routing protocol
    (LBRP), 209
Location-based services (LBS), 136
Long Range (LoRa), 141, 170
Long short-term memory (LSTM), 112, 115
Long-term evolution (LTE), 208
LoRaWAN, 59, 141, 143
Low-power wide area network (LPWAN), 137
Low-power wireless personal area networks
    (LoWPANs), 151, 224

M

Machine-to-machine (M2M), 108, 134, 172
Maximum power point tracker
    (MPPT), 226
Mesh, 17, 48, 60, 155
Message Query Telemetry Transport (MQTT),
    58, 155
Metal oxide semiconductor (MOS), 82
Micro controller unit (MCU), 228, 246
MQ sensors, 258

N

Narrow Band IoT (NB-IoT), 60, 138
Near field communication (NFC), 59, 147, 147, 169
Network layer, 45, 151

O

Open Systems Interconnection (OSI), 21
Output-stationary (OS) dataflow, 119

P

Perception layer, 45
Point-to-point, 15, 57
Pulse width modulation (PWM) 261

Q

Quality of service (QoS), 14, 155

R

Radio frequency identification (RFID) 146, 169
Reconfiguration 9
Reduced instruction set computer (RISC) 260
Ring 17
Road side unit (RSU), 213
RS-232, 21
RS-422, 22
RS-485, 23, 24

S

Scalability, 4, 6, 58, 115, 121
Security, 45, 50, 108, 112, 154
Sense amplifiers, 82
Sensor module, 228
Sensors, 28, 224, 257
Serial peripheral interface (SPI), 76
SigFox, 59, 144

Single instruction multiple data (SIMD), 188
Smart agriculture, 8, 143, 169, 171
Smart city, 162, 164
Smart healthcare, 8
Smart home, 8, 17, 28, 108, 191
Smart industry, 9
Social IoT, 49
Static random-access memory cell (SRAMC), 76

T

TIA/EIA 232, 21
TIA/EIA 422, 22
TIA/EIA 485, 23

U

User datagram protocol (UDP), 171

V

Vehicle-to-cloud (V2C), 212
Vehicle-to-infrastructure (V2I), 212
Vehicle-to-vehicle (V2V), 211
Vehicular ad-hoc networks, 208
Very large-scale integrated circuit (VLSI), 76, 174

W

Weight-stationary dataflow, 118
Wireless, 25, 36, 57, 139
Wireless sensor networks (WSN), 169, 207
Wireless-IEEE 802.11, 20, 57, 72
Write enable (WE), 81

Z

Zigbee, 58, 65, 150, 170
ZigBee, 150, 170
Z-wave, 59, 152